Fuzzy Logic
and
Hydrological Modeling

Zekâi Şen

CRC Press
Taylor & Francis Group
Boca Raton   London   New York

CRC Press is an imprint of the
Taylor & Francis Group, an **informa** business

CRC Press
Taylor & Francis Group
6000 Broken Sound Parkway NW, Suite 300
Boca Raton, FL 33487-2742

First issued in paperback 2017

© 2010 by Taylor and Francis Group, LLC
CRC Press is an imprint of Taylor & Francis Group, an Informa business

No claim to original U.S. Government works

ISBN 13: 978-1-138-11355-8 (pbk)
ISBN 13: 978-1-4398-0939-6 (hbk)

This book contains information obtained from authentic and highly regarded sources. Reasonable efforts have been made to publish reliable data and information, but the author and publisher cannot assume responsibility for the validity of all materials or the consequences of their use. The authors and publishers have attempted to trace the copyright holders of all material reproduced in this publication and apologize to copyright holders if permission to publish in this form has not been obtained. If any copyright material has not been acknowledged please write and let us know so we may rectify in any future reprint.

Except as permitted under U.S. Copyright Law, no part of this book may be reprinted, reproduced, transmitted, or utilized in any form by any electronic, mechanical, or other means, now known or hereafter invented, including photocopying, microfilming, and recording, or in any information storage or retrieval system, without written permission from the publishers.

For permission to photocopy or use material electronically from this work, please access www.copyright. com (http://www.copyright.com/) or contact the Copyright Clearance Center, Inc. (CCC), 222 Rosewood Drive, Danvers, MA 01923, 978-750-8400. CCC is a not-for-profit organization that provides licenses and registration for a variety of users. For organizations that have been granted a photocopy license by the CCC, a separate system of payment has been arranged.

Trademark Notice: Product or corporate names may be trademarks or registered trademarks, and are used only for identification and explanation without intent to infringe.

Library of Congress Cataloging-in-Publication Data

Sen, Zekâi.
    Fuzzy logic and hydrological modeling / Zekâi Sen.
       p. cm.
    Includes bibliographical references and index.
    ISBN 978-1-4398-0939-6 (alk. paper)
    1. Hydrologic models--Mathematics. 2. Fuzzy logic. 3. Spatial analysis (Statistics) I. Title.

    GB656.2.H9S46 2009
    553.701'1--dc22
                                       2009015354

Visit the Taylor & Francis Web site at
http://www.taylorandfrancis.com

and the CRC Press Web site at
http://www.crcpress.com

Dedication

Bismillahirrahmanirrahim

In the name of Allah, the most Merciful and the most Beneficial.

Contents

Preface

Water resource system planning, operation, and maintenance require different sources of information, including meteorology, hydrology, geology, limnology, and computer programming and management strategies. Hydrological sciences require temporal and spatial data sources for a proper understanding of the phenomena concerned. For this purpose, all over the world there are monitoring stations of different hydrology, hydrometeorology, meteorology, and hydrogeology variables, and water quality measurement instruments for numerical data records. In any study, prior to anything else, the water scientist seeks available reports and especially numerical data related to his or her problem. Although this is necessary it is not sufficient because prior to any study there is logic- and experience-based (expert views) information available. It is this information that provides the foundation for the preparation and further interpretation and deduction of logically acceptable conclusions. There are problems for which the readily available equations and algorithms may not be enough for solution. However, basic linguistic knowledge, experience, and expert views help establish the preliminary skeleton of the solution, which may later be supported by the numerical data. Unfortunately, so far in the literature, numerical data are pumped into mathematical models, especially through readily available computer software, which may produce unreliable results if the background of the working mechanism related to any natural water phenomenon is not appreciated qualitatively through verbal information.

Classical hydrology problem solution approaches work with crisp and organized numerical data on the basis of two-valued (white-black, on-off, yes-no) crisp logic. In almost every corner, hydrological sciences have gray fore- and backgrounds with verbal information. The big dilemma is how to deal with gray information for arriving at decisive conclusions with crisp and deterministic principles. Fuzzy logic (FL) principles with linguistically valid propositions and vague categorization provide a sound ground for the evaluation of such information. The preliminary step is FL conceptualization of the hydrological sciences phenomena with uncertainties in its input, system, and output variables. Such an approach helps not only to visualize the relationships between different variables, but also furnishes philosophical, logical, and rational details about the system mechanism of the hydrological sciences phenomena without mathematical formulation.

The main purpose of this book is to prepare the reader for linguistic (verbal), qualitative data interpretation and treatment procedures on the basis of FL principles. The reader should always keep in mind any data related to natural events such as water-related phenomena, including vagueness, imprecision, incompleteness, and missing data, which are collectively referred to as "fuzzy" information. This information is a kind of uncertainty—not only numerically, but also linguistically. For instance, when somebody is given the information that "some hydrological variable is low," he/she tries first to interpret this verbal expression qualitatively and then quantitatively. However, this quantification will not be a crisp number, but rather an uncertain variation interval. Furthermore, the hydrological variable within such an interval will not be accorded the same significance or value of truth. This is the rough definition of fuzzy values, and then the question is how to process such fuzzy data. The fuzzy sets, systems, models, reasoning, and inference are explained in this book with simple examples. Provided that the basic fuzzy information principles and processes are understood properly, they can be employed effectively in any problem in hydrological sciences.

FL systems use information efficiently; all available evidence can be used until the final defuzzification, which may be reduced to a crisp solution that is robust to uncertainty, missing data, or vague information. They encode human expert knowledge/heuristics and common sense under naturally enforced constraints in a simple and cheap manner without any need for training data, mathematical formulations, or probability distributions functions. Their designs are relatively straightforward and implementations are stimulating. Designing the fuzzy sets—and hence relevant logical rules—for the hydrological phenomena generation mechanism is comparatively easier than any other type of modeling. Fuzzy systems are stable, easily tuned, and can be conventionally validated.

If a real-world problem is sought that is inherently messy, where mathematical functions are difficult to apply, fuzzy modeling of hydrological processes would be an excellent example. Hydrologic processes depend on topography, vegetation, soil moisture, rainfall pattern and intensity, potential evapo-transpiration, air temperature, solar radiation, wind, and dew point, factors that change spatio-temporally in an imprecise, vague, and uncertain manner. Nonetheless, it is necessary to calculate hydrologic processes in this real-world environment. Hydrologic models are useful only to the degree that they represent processes in the real world.

This book is designed in such a manner that the first seven chapters expose the FL principles, processes, and design for a fruitful inference system with many hydrological examples. The final two chapters present the use of FL principles on larger hydrological scales within the hydrological cycle.

I do not know how to express my spiritual feelings toward my wife Fatma Şen, for her patience, unlimited understanding, and continuous support during the writing of this book. I pray for her from the bottom of my heart because she gave me, after Allah, this golden opportunity to serve the whole of humanity with my worldly science. The flavor and satisfaction of science are in its service to all humanity.

Zekâi Şen
Istanbul, Turkey

About the Author

Zekâi Şen obtained both a B.Sc. and an M.Sc. from the Technical University of İstanbul in 1971, and continued his post-graduate studies at the University of London, Imperial College of Science and Technology. He obtained a Diploma of Imperial College (DIC) and M.Sc. in engineering hydrology in 1972, and Ph.D. in stochastic hydrology in 1974. He has published nearly 400 scientific papers in more than 50 different top international journals on various topics. Dr Şen has also supervised numerous international M.Sc. and Ph.D. degree students, and holds several national and international scientific prizes including a Nobel Peace Prize presented for his contribution to the joint Intergovernmental Panel on Climate Change (IPCC 2007).

Dr Şen has written the following books in English: *Applied Hydrogeology for Scientists and Engineers*, *Wadi Hydrology*, *Solar Energy Fundamentals and Modelling Techniques (Atmosphere, Environment, Climate Change and Renewable Energy)*, and *Spatial Modeling in Earth Sciences*. In addition, Dr Şen has also written many books in Turkish, and one book in Arabic. He is a member of the American Institute of Hydrology, and is also the president of Turkish Water Foundation.

Dr Şen is currently a faculty member in the civil engineering department at the Technical University of Istanbul. His main interests are hydrology, water resources, hydrogeology, hydrometeorology, hydraulics, science philosophy and history.

1 Introduction

1.1 GENERAL

Hydrologic events in nature are complex and arise from uncertainty in the forms of ambiguity. Subconscious human thinking digests complexity and ambiguity with partial solutions in natural, earth, and social events. Although the complete description of a real hydrologic phenomenon often requires detailed data, human perceptions and reasoning economize this requirement by pondering the generating mechanism of the phenomenon concerned (Chapter 2). This is due to the fact that humans have the capability of approximate reasoning about the behavior of the phenomenon, which leads to a generic understanding of the problem. As Zadeh (1973) states:

> "Complexity and ambiguity are related, the closer one looks at a real world problem, the fuzzier becomes its solution"

The *Oxford English Dictionary* defines the word *fuzzy* as "blurred, indistinct, imprecisely defined, confused, vague," which gives the impression that there is no use of this word in daily life except with dangers, and outside of science and technology. However, the real situation is very opposite to such thinking. Fuzzy logic (FL) and system design have attracted the growing interest of researchers and various scientific and engineering areas. FL is based on the way the brain deals with inexact information. Specifically, FL systems take the linguistic and numerical information as dual data sources and provide the answer accordingly in a vague manner, which includes numerous solutions. It is better to start with FL principles and arrive at a set of fuzzy conclusions than to conclude with classical logic (CL) (two-valued logic) a mathematical approach with only one crisp result, which may never appear in real life.

Learning through observation, measurement, experience, and reasoning about complex phenomena reduces the fuzziness in understanding but never eliminates it. A reduction in the complexity means exploration of the problem with more certain ideas. As the complexity becomes marginal with little uncertainty, mathematical formulations provide a precise description of the generating mechanism. For instance, probabilistic, statistical, and stochastic modeling in the hydrological sciences becomes applicable for the description and even control of the phenomenon. Most often, in many branches of science, mathematical models are based on a set of assumptions, idealizations, and simplifications in order to reduce complexity in an artificial manner so as to establish at least preliminary approximations that can be expressible by formulations, which are necessary to make quantitative and deterministic conclusions. Mathematical formulations are necessary in finding numerical solutions for the problem at hand. Any mathematical model is based on a set of restrictive assumptions, which overlooks the fuzziness in the problem.

Hydrological events are too complicated for precise descriptions, and therefore approximate reasoning (fuzzy) must be introduced to obtain reasonable yet tractable models. Deterministic (crisp) models ignore fuzzy human knowledge in stochastic and mathematical approaches. In the hydrology domain, it is most needed to depend on a model, which can digest and formulate human knowledge in a systematic manner and can integrate it with other information sources such as field measurements and linguistic information.

Many hydrological processes have nonlinear behavior but due to mathematical requirements and restrictions, great effort has been put into linear systems in the past. In a way, the complexity of the involved mathematics directed the research toward the linear pole. In general, a good model is expected to be precise to the extent that it characterizes the key features of the hydrologic phenomena and, in the meantime, it is tractable for mathematical analysis. The branch of FL modeling is completely independent of mathematical principles and formulations, all of which have logical rule bases in their roots. FL systems and models are capable of digesting linguistic and numerical information, whereas mathematical models are based on numerical information only. In particular, as the worth of linguistic (verbal) information became increasingly appreciated in recent decades—for example, "know-how"—the use of FL methodology became more attractive.

In hydrological systems, valuable information has two sources: hydrologist's expert views (Chapter 2) and measurements from various instruments. In classical hydrologic modeling, no one really cared about the former source of information until recently. Even a shepherd may have efficient information, for example, about the groundwater resources in and around his village. In classical techniques, there is no role for his valuable linguistic information but FL approaches put this information into the internal structure of modeling through fuzzy sets (Chapter 3). Herein, the key question is how to transform an expert hydrologist's knowledge first into a logical rule base and then into a mathematical framework or formula. Essentially, this is the achievement of FL system design and inference (Chapters 6 and 7).

The concept of fuzzy sets was introduced by Zadeh (1965), who pioneered the development of FL instead of CL. Unfortunately, this concept was not welcome in the literature as many uncertainty techniques (such as the probability theory, statistics, and stochastic processes) were commonly employed at that time; however, FL has been developing since then and is now being used especially in Japan for automatic control in commercially available products. Many textbooks provide basic information on the concepts and operational fuzzy algorithms (Dubois and Prade, 1988; Kosko, 1987; Ross, 1995). The key idea in FL is the allowance of partial belonging of any object to different subsets of the universal set instead of belonging to a single set completely.

Fuzzy applications in hydrology and meteorology are rather rare. Kindler (1992) used FL for optimal water allocation. Bardossy and Disse (1993) applied it to model the infiltration and water movement in the unsaturated zone. Russel and Campbell

(1996) studied fuzzy rule-based control systems for reservoir operation. Aronica et al. (1998) used FL in calibrating distributed roughness coefficients in a flood propagation model with limited data. Şen (1998) applied-fuzzy algorithm to estimate solar irradiation from sunshine duration. Pongracz et al. (1999) found that fuzzy rule-based methodology on regional drought provided an excellent tool. Tayfur et al. (2003) applied an FL algorithm to predict the mean sediment load from bare soil surfaces subjected to rainfall/runoff-driven sediment transport.

FL is a powerful problem-solving methodology with a myriad of applications in embedded control and information processing. Its principles provide a remarkably simple way to draw definite conclusions from vague, ambiguous, or imprecise information. In a sense, FL resembles human decision making with its ability to work from approximate data and find precise solutions.

It is emphasized in this book that the FL approach can provide the structure and solution procedure of hydrologic systems prior to any crisp (deterministic) method such as mathematics, statistics, or probability or stochastic processes. In this way, a hydrologist is able to develop creative and analytical thinking capabilities with the support of other expert views. Because the modern philosophy of science insists on the falsification of current scientific results (Popper, 2002), there is always room for ambiguity, vagueness, imprecision, and fuzziness in any scientific research activity. FL attributes degrees as scientific beliefs (degree of verification or falsification) that assume values between 0 and 1, inclusively (Chapter 6). The verifiability of scientific knowledge or theories by a logical positivist means, on classical grounds, that the demarcation of science concerning a phenomenon is equal to 1 without allowing room for falsification. The conflict between verifiability and falsifiability of scientific theories includes philosophical grounds that are fuzzy. Although many science philosophers tried to resolve this problem by bringing into the argument the probability and at times the possibility of the scientific knowledge demarcation and development, unfortunately so far the "fuzzy philosophy of science" has not been introduced into the hydrology literature (Şen, 2007). Hence, it is the purpose of this book to give an account of FL in the treatment of hydrologic knowledge through appropriate fuzzy models. The dogmatic nature of scientific knowledge without a doubt may arise from classical logic (CL) with its dual crisp values as black (one) and white (zero), ignoring gray tones, whereas FL holds the scientific arena vivid and fruitful for future plantations and knowledge generation. Innovative hydrological systems should lean more toward the basic scientific philosophy of problem solving with FL principles. Contrary to the CL, FL may be thought to be basically similar to a multi-valued logic. However, this is not exactly so, due to uncertain boundaries between the multiple subsets. It allows intermediate uncertain values to be defined between two-valued conventional evaluations like "dry"/"wet," "high"/"low," "intense"/"sparse," etc. Notions like "rather dry," "highly humid," or "semi-arid" cannot be formulated crisply except through FL systems and models.

This book presents systematic and comprehensive modeling of uncertainty, vagueness, or imprecision through FL principles, procedures, and systems for problem solving in hydrological sciences. There are several chapters that introduce FL basic definitions, fuzzy sets, clustering, rule base, and inference systems, and subsequent chapters that provide hydrological applications (Chapters 8 and 9).

1.2 FUZZINESS IN HYDROLOGY

Hydrology problem solutions include subjective linguistic information, which is frequently hard to quantify using CL methodologies such as mathematics, probability theory, stochastic processes, and statistics. These methodologies require a set of assumptions that leads to solutions, where the linguistic information is ignored. Wide ranges of water engineering and hydrological parameters are characterized by uncertainty, subjectivity, imprecision, and ambiguity. For instance, human operators of water resources systems, say dams, use subjective knowledge or linguistic information on a daily basis in making decisions for leaving a readily available empty volume in reservoir storage for the next flood, in saving water during wet periods to offset drought duration, and in calculating design flood and peak discharges. The environment in which a hydrologist makes decisions is most often complex, which makes it difficult to formulate a suitable mathematical model.

To cope with complex situations, the FL principles and modeling are justified because such modeling digests linguistic information in addition to numerical data. In many situations, linguistic information is obtainable through observations more easily than through numerical information. In any preliminary hydrologic field study, there may not be numerical data, but the hydrologist's observations provide a set of linguistic information that leads to logical and rational thinking with preliminary approximate deductions and solution rules (Chapter 6). For instance, water taste by tongue gives expert information about the quality, or looking at the rock outcrop provides the first impression about the infiltration rate and groundwater recharge. The historical traces of flood water level on both sides of a cross-section provide qualitative information about past flood discharges. In solving real-life hydrology problems, hydrologists should use not only objective information (equations, algorithms, and formulations) or only subjective knowledge (linguistic information), but one should also exploit both information sources in arriving at an optimum solution. FL principles are extremely suitable for combining linguistic knowledge with objective information (Chapter 7).

FL has gained increasing acceptance during the past few decades. Nearly every application can potentially realize some of the FL benefits, such as performance, simplicity, lower cost, and productivity. FL has been found very suitable for embedded control applications (Dubois and Prade, 1988; Kandel and Langholz, 1994; Klir and Folger, 1988; Ross, 1995). Almost everybody has had some contact with CL at several points in their lives, where a statement is either "true" or "false," with nothing in between. This principle of either "true" or "false" was formulated by Aristotle some 2300 years ago as the Law of the Excluded Middle, and has dominated Western logic ever since. Of course, the idea that things must be either "true" or "false" is in many cases nonsense. Is the statement "The weather is good" completely "true" or completely "false"? Probably neither of them is "true" or "false." How about the rainfall intensity? And how about most of the drainage area? Or the recharge is "poor?" The idea of granulations of truth is familiar to everyone.

On the other hand, for most complex systems with few numerical data where only ambiguous and imprecise information is available, none of the deterministic formulations in hydrological sciences help in solving the problem. However, fuzzy reasoning provides a way to understand and then interpret system behavior with interpolations based on the available scarce numerical but rather abundant verbal (linguistic) data about the generating mechanism of the phenomenon with its inputs (causes, antecedents) and outputs (results, consequents). In fuzzy reasoning, the information about the input and output variables is combined with logical insights about the system. This provides an ability to describe, in words through a set of rules, the mathematical abstraction of the real world (Chapter 6). Fuzzy modeling includes first matching ambiguous input and output information through FL rules and then inferring results through fuzzy inference system (FIS) modeling (Chapter 7). Such a procedure requires reasoning with logical footprints that constitute the backbone of the behavioral abstraction of the problem with rational and partial conclusions. Due to complexity and ambiguity, human ability provides inference by reasoning the internal mechanism of the problem, which requires not only crisp conceptions and mathematical formulations, but also more artistic scenarios of different alternatives. A detailed explanation of fuzzy modeling in engineering aspects was already presented by Ross (1995) and the necessary fundamentals are presented in Chapters 2, 3, and 4 in this book.

Classical water resources systems work with crisp and organized numerical data on the basis of CL, which has only two mutually exclusive alternatives like wet (white, yes, one, positive, true, etc.) and dry (black, no, zero, negative, false, etc.). Hydrological sciences have gray fore- and backgrounds with verbal information in almost every corner, which are full of ambiguous, vague, imprecise, and generally uncertain information sources. It is a big dilemma as to how to deal with gray information in arriving at decisive conclusions with crisp and deterministic principles. However, FL principles with linguistic premises and vague categorization provide sound ground for the digestion of such information. The preliminary step is FL conceptualization of a hydrologic phenomenon with uncertainties in its input, system, and output variables (Chapters 6 and 7). Such an approach helps not only visualize the relationships between different variables, but also furnishes philosophical details about the system mechanism of the hydrological phenomenon without mathematical formulation.

Fuzzy set theories provide a rich and meaningful addition to two-valued CL (Chapter 3). The mathematics generated by these theories is consistent, and FL seems to be the generalization of CL. The applications that may be generated from or adapted to FL are wide-ranging and provide the opportunity to model conditions that are inherently imprecise despite the concerns of classical logicians. Many hydrologic systems can be modeled, simulated, and even replicated with the help of FL systems, not the least of which is human reasoning itself (see Chapters 2 and 5).

Unlike CL, which requires a deep understanding of a system, exact equations, and precise numeric values, FL incorporates an alternative way of thinking that allows modeling complex systems using a higher level of abstraction originating from our knowledge and experience. FL allows expression of this knowledge with subjective concepts such as "very hot," "bright red," and a "long time," which are mapped into exact numerical ranges (see Chapter 3).

FL offers a better way of representing reality. In FL, a statement is "true" to various degrees—ranging from "completely true," through "half-true" or "half-false," to "completely false." The basic idea of multi-valued logic has been explored to some extent by a number of mathematicians in this century. Zadeh (1965) published the first articles on the theory of fuzzy sets, which gave rise to hundreds of articles on fuzzy mathematics and system theory. Most importantly, Japanese advances in the field of fuzzy control have won the attention of engineers throughout the world. Fuzzy theory provides a whole new approach to the mathematics of thinking (Chapter 2).

FL-based research in hydrology is still in its early stages. A great deal of empirical experimentation in FL hydrological modeling applications will be required before these approaches are established as common methodologies, especially among those hydrologists who would appear to have moved away from deductive (top-down), crisp, black-box models, in favor of more physically based inductive (bottom-up) gray models.

There is an overwhelming amount of uncertainty associated with hydrological complex systems and it is non-random in nature. For instance, "drought," "flood," "dry," and "wet" are four commonly used concepts in hydrology that include linguistic uncertainty, which becomes vaguer when one says "severe flood, "very dry," "almost wet," "intensive drought," "slightly wet," etc. Fuzzy set theory helps in dealing with this type of vagueness in modeling the natural events with imprecision and/or a lack of information regarding the problem at hand (Chapter 3). In any hydrological modeling, most often one understands that there is some lack of complete information in the solution. Some of the information may be judgmental and qualitative in words. All these can be incorporated in the FL modeling of hydrological processes in addition to perceptions about the phenomenon. Different steps of fuzzy modeling are discussed in the chapters of this book.

Human common sense is either applied from what seems reasonable, for a new system, or from experience for a system that has previously had a human operator. Here is an example of converting human experience for use in a hydrologic system. Water engineers are not able to automate with conventional logic. Eventually, they translate the human "perception" into lots of fuzzy "IF . . . THEN . . ." rules based on human experience. Reasonable success was thereby obtained in automating the plant. Objects of FL analysis and control may include physical control, such as flow speed or operating a dam, financial and economic decisions, psychological conditions, physiological conditions, safety conditions, security conditions, and much more.

Human beings have the ability to take in and evaluate all sorts of information from the physical world with which they are in contact and to mentally analyze, average, and summarize all these input data into an optimum course of action. All living things do this, but humans do it more and do it better with fuzzy conceptions (Chapter 2).

An obvious drawback to FL is that it is not always accurate. The results are perceived as a guess, so it may not be as widely trusted as would an answer from CL. Certainly, however, some chances must be taken. How else can hydrologists succeed in modeling by assuming that the runoff coefficient is 0.35?

The degree of fuzziness of a system analysis rule can vary between being very precise, in which case one would not call it fuzzy, to being based on an opinion held by a human, which would be fuzzy. Being fuzzy or not fuzzy, therefore, has to do with the degree of precision of a system analysis rule. A hydrologic system analysis rule can be based on human fuzzy perceptions under the light of incomplete and vague information and personal experience even though it may be subjective. For example, one could state as a rule:

"IF the reservoir level rises to a danger point, say, 1 m below the spillway crest, THEN open the gates."

This rule is not fuzzy because "open the gates" is a crisp expression. However, its fuzzy counterpart can be expressed as:

"IF the reservoir level rises to a danger point, say, about 1 m below the spillway crest, THEN starts to open the gates slightly."

The addition of two words, *about* and *slightly*, renders the crisp sentence a fuzzy statement.

It is clear today that description and generation leading to satisfactory mathematical structure of any physical actuality are often unrealistic requirements.

> In any axiomatic mathematical system (theory), there are fuzzy propositions, that is, propositions which cannot be proved or disproved within the axioms of this system.
> **—Gödel, 1932**

> All traditional logic habitually assumes that precise symbols are being employed. It is, therefore, not applicable to this terrestrial life but only to an imagined celestial existence.
> **—Russel, 1948**

> So far as the laws of mathematics refer to reality, they are not certain. And so far as they are certain, they do not refer to reality
> **—Albert Einstein**
> *Quoted in J.R. Newman, The World of Mathematics, New York, 1956*

> As the complexity of a system increases, our ability to make precise and yet significant statements about its behavior diminishes until a threshold is reached beyond which precision and significance (or relevance) become almost mutually exclusive characteristics.
> **—Zadeh, 1965**

1.3 WHY USE FUZZY LOGIC IN WATER SCIENCES?

A fundamental question is, "When is it appropriate to use FL?" The answer is, "When the process is concerned with continuous phenomena (one or more input variables are continuous) that are not easily broken down into discrete segments; when a mathematical model of the process does not exist, or exists but it is difficult to encode, or it is too complex to be evaluated fast enough for real-time operation; when the process involves human interaction (such as human descriptive or intuitive thinking); and when an expert is available who can specify the rules underlying the system behavior and the fuzzy sets that represent the characteristics of each variable. FL finds its applications in such areas as:

1. Natural language processing by words and sentence calculations
2. Inference between reasons and results concerning natural phenomena such as hydrologic processes
3. Expert systems without formulations
4. Planning in various activities
5. Prediction of mechanical processes
6. Estimation of future events
7. Control processes
8. Pattern recognition, classification, and image processing
9. Operation and management of various systems such as water resources systems
10. Information retrieval

Present hydrology systems are rather classical, with extensive dependence on crisp and blueprint type of information. In many institutions, knowledge and information are spoon-fed to fresh minds without any creative or functional productivity. This is perhaps one of the main reasons why in many institutions all over the world, creative and analytical thinking capabilities are not advanced. In classical systems, more than basic logical propositions, formulations, and determinism are mentioned for the solution of problems. In particular, in hydrological sciences, almost each field trip to a site is a rather different case study from other sites although they may be geographically close to each other. Therefore, CL or deterministic information systems cannot be fully valid for the description of the hydrological phenomena concerned.

Hydrologic knowledge cannot be verifiable completely or falsifiable, but rather fuzzifiable, which provides the potential for further research. It is asserted in this book that hydrological sciences will not be completely verifiably or falsifiable but will always be fuzzifiable and hence further developments in the form of prescience, traditional science, and occasional revolutionary science will be in view for all times, spaces, and societies (Kuhn, 2000).

Hydrological sciences are full of ambiguous, vague, imprecise, and, in general, uncertain information sources that can be treated with the FL concepts for clear ideas and solutions. There are different versions of hydrology, such as deterministic hydrology, stochastic hydrology, watershed hydrology, wadi hydrology, contamination hydrology, etc., which require almost certain information. However, if there are random uncertainties, then they are accounted for by the classical uncertainty techniques such as probability, statistics, or stochastic processes and more recently by chaos theory and similar approaches. The hydrological models are mathematically based on two CL alternatives as the Law of the Excluded Middle. Accordingly, mathematical equations, systematic algorithms, and formulations are the basis of the modeling for estimation, prediction, model identification, or filtering purposes. In probabilistic and stochastic modeling of hydrological processes, a set of assumptions is necessary, such as the stationarity, homogeneity, ergodicity, intrinsicity, etc. These assumptions complex hydrologic phenomena to manageable classical mathematical sizes and domains. Otherwise, the ignorance of the hydrologist cannot be accounted for by CL, which constitutes the foundation of mathematical models, the success of which depends on the numerical database. This is the reason why every hydrology unit in the world would have to have a sound database. On the other hand, FL furnishes the premises of the rule base first.

Fuzziness is often confused with probability; whereas the former deals with deterministic plausibility, the latter concerns the likelihood of nondeterministic stochastic events. Fuzziness (vagueness) found in the definition of a concept or the meaning of a term such as *semi-pervious* aquifer, *long time*, or *moderate distance*. It conveys suspective human thinking, feeling, or language (Chapter 2). However, the uncertainty of probability generally relates to the occurrence of phenomena as symbolized by the concept of randomness. For example, statements such as "It will rain tomorrow" and "There is a sandstone layer at 50 meter depth" have the uncertainty of random character. From the modeling point of view, fuzzy models and probabilistic, statistical, or stochastic models possess different philosophical information. FL has fuzzy sets and hence membership functions (MFs,) which represent similarities of objects to imprecisely defined properties, while probabilistic, statistical, and stochastic models convey information about relative frequencies that are based on crisp sets and on CL.

Water-related earth sciences—hydrology, hydrometeorology, hydrometry, hydrochemistry, etc.—are concerned mainly with natural phenomena and involve uncertainties. Among these phenomena are rainfall, runoff, flood, drought, thunderstorm, fog, solar irradiation, evaporation, wind velocity, etc., which are closely related to daily social and economic activities. It is necessary to predict future occurrences and magnitudes of these phenomena with a certain reliability in order to control disastrous consequences for society. Successful prediction models are based on the quantitative uncertainty methods of probability, statistics, and stochastic processes. In this book, FL-based estimation and simulation models are presented in the domain of fuzzy hydrology.

Fuzzy hydrology incorporates in its model construction, processing, and control stages, non-random (linguistic) uncertainties together with the numerical database. The fundamental skeleton of fuzzy modeling is the rule base rather than the database. The rule base includes all the linguistic data in the form of fuzzy sets with MFs, which are communicators between computers and human mentality (Chapter 3). Hence, fuzzy hydrology can be defined as a new hydrological research alternative where suitable model identification for the problem at hand has logical rules with fuzzy sets as basic ingredients in making relevant connective statements between the input and output variables of the system. There are two basic types of uncertainty that may be present in any real-world hydrologic process:

1. *Stochastic uncertainty:* This is due to a lack of information, where the future state of the system may not be known completely. It has been handled by probability theory, statistics, and stochastic processes. The outcome of a stochastic event is either *true* or *false.*
2. *Fuzziness:* This is vagueness concerning the description of the semantic meaning of the events, phenomena, or statements themselves. In this situation, where the event itself is not well defined, the outcome may be given by a quantity other than true (one) or false (zero). That is, the outcome in the

presence of fuzziness may be quantified by a *degree of belief* (Chapter 6). The events are modeled as fuzzy sets (Chapter 3) because the characteristic functions (MFs) of such sets may take values other than zero or one.

Although probabilistic, statistical, and stochastic approaches and methods have been used for many years in hydrology, linguistic knowledge still could not be digested. Inclusion of linguistic data in hydrologic systems brings additional dimension to problem solving.

The decision-making process in hydrology must necessarily deal with imprecise information. The information, such as completion of a project, for instance, would not be precisely known before it is finished. The volume of a riverbank dredging would also be only an imprecise estimation. This type of information is not probabilistic; therefore, it cannot be handled appropriately with probabilistic formality. Rather, it should be handled with fuzzy hydrology principles.

REFERENCES

Aronica, G., Hankin, B., and Beven, K. 1998. Uncertainty and equi-finality in calibrating distributed roughness coefficients in a flood propagation model with limited data. *Adv. Water Resour.* 22(4): 349–365.

Bardossy, A. and Disse, M. 1993. Fuzzy rule based models for infiltration. *Water Resour. Res.* 29(2): 373–382.

Dubois, D. and Prade, H. 1988. *Possibility Theory: An Approach to Computerized Processing of Uncertainty.* Plenum Press, New York.

Gödel, K. 1932. Introductory note to 1932 A. S. Troelstra Zum intuition istischen assagentoallrül (on the intuitionistic propsitional calculus.)

Kandel, A. and Langholz, G. 1994. *Fuzzy Control Systems.* CRC Press, Boca Raton, FL.

Kindler, J. 1992. Rationalizing water requirements with aid of fuzzy allocation model. *J. Water Resour. Plann. Manag.* 118(3): 308–323.

Klir, G.J. and Folger, T. 1988. *Fuzzy Sets, Uncertainty and Information.* Prentice Hall, Englewood Cliffs, NJ.

Kosko, B.1987. *Fuzzy Thinking. The New Science of Fuzzy Logic,* Hyperion, New York.

Kuhn, T. 2000. *The Structure of Scientific Revolutions.* University of Chicago Press, Chicago.

Pongracz, R., Begird, I., and Duckstein, L. 1999. Application of fuzzy rule-based modeling technique to regional drought. *J. Hydrol.* 224: 100–114.

Popper, K. 2002. *The Logic of Scientific Discovery.* Routledge, London.

Ross, J.T. 1995. *Fuzzy Logic with Engineering Applications.* McGraw-Hill, New York.

Russel, B. 1948. *Human Knowledge: Its Scope and Limits.* London, George Allen and Unwin.

Russel, S.O. and Cambell, P.E. 1996. Reservoir operating rules with fuzzy logic programming. *J. Water Resour. Plann. Manag.* 122(4): 262–269.

Şen, Z. 1998. Fuzzy algorithm for estimation of solar irradiation from sunshine duration, *Solar Energy* 63(1): 39–49.

Şen, Z. 2007. *Fuzziology-Fuzzy Philosophy of Science and Education. Critical Review.* A Publication of Society for Mathematics of Uncertainty (Ed., Paul P. Wang), pp. 33–44.

Tayfur, G., Özdemir, S., and Singh, V.P. 2003. Fuzzy logic algorithm for runoff-induced sediment transport from bare soil surfaces. *Adv. Water Resour.* 26: 1249–1256.

Zadeh, L.A. 1965. Fuzzy sets. *Information Control* 8: 338–353.

Zadeh, L.A. 1973. Outline of a new approach to the analysis of complex systems and decision processes. *IEEE Trans. Syst. Man. Cybern.* 3: 28–44.

PROBLEMS

1.1 Which of the following words imply fuzziness in hydrological studies?
   (a) "complexity"
   (b) "ergodicity"
   (c) "ambiguity"
   (d) "stationarity"
   (e) "ensemble"

1.2 Which of the following hydrological studies can be regarded as a probabilistic, statistical, or stochastic modeling issue?
   (a) "rather dry periods"
   (b) "dry durations"
   (c) "main stream length"
   (d) "cloudy weather"
   (e) "soil types"

1.3 Which of the following deeds derive fuzzy thinking?
   (a) "imagination"
   (b) "philosophical thinking"
   (c) "set of assumptions"
   (d) "energy conservation principle"
   (e) "description"

1.4 Order the following hydrological tasks by degree of uncertainty:
   (a) well hydraulics
   (b) wadi hydrology
   (c) water resources management
   (d) infiltration
   (e) water supply

1.5 Do all the following statements include fuzziness?
   (a) There is a 90% chance of rainfall tomorrow.
   (b) If it rains then there will be runoff.
   (c) Reservoir operation needs expert views.
   (d) Volume of water per time interval is defined as discharge.
   (e) Conceptual models are useful in describing simple behaviors of a hydrologic phenomenon.

1.6 Identify fuzzy statements among the following alternatives:
   (a) There is the possibility of flooding in January.
   (b) Synthetic sequences are statistically indistinguishable from the historical records.
   (c) Rational formula gives the value of peak discharge after each storm rainfall.
   (d) The Darcy law in groundwater relates hydraulic gradient to seepage velocity through hydraulic conductivity.
   (e) Past records are the reflection of the same phenomenon in the future.

1.7 Which of the following need fuzzy logic?
   (a) Daily dry and wet spells occur in a time sequence.
   (b) A shepherd's explanation of the rainfall phenomenon in his village.
   (c) Software output of storm hydrograph.
   (d) $Q = AV$, where Q, A, and V are the discharge, cross-sectional area, and velocity, respectively.
   (e) Preliminary field trip well inventory.

1.8 Identify fuzzy imprints in the following statements, and criticize if necessary.
   (a) Human expert knowledge plays a significant role in scientific developments.
   (b) Innovators work with two-valued logic.

(c) In analytical derivations, scientists use crisp logic.

(d) Fuzzy logic is the backbone of scientific, technological, and innovative developments.

(e) Human communication needs language media and it includes fuzzy logic principles more than crisp logic.

1.9 Can fuzzy logic principles be used in all the following investigations?

(a) The more the rainfall, the more the infiltration.

(b) Discharge is directly related to rainfall intensity.

(c) Water resources operation and management.

(d) Hydrological knowledge cannot be completely verifiable or falsifiable.

(e) Mature hydrological information can be obtained by deterministic approaches over time.

1.10 What are the fuzzy logic features in the following hydrological sentences?

(a) Authorized knowledge with precautions and the solutions.

(b) Hydrological processes include uncertainty and suspicion, which can be dealt with using probabilistic methodologies.

(c) Philosophical thinking must be trimmed by two-valued logic to arrive at mathematical formulations by a set of assumptions, idealizations, and simplifications.

(d) All hydrological formulations are falsifiable and hence they can be improved for a better formulation.

(e) Conceptual models have fuzzy statements in addition to two-valued logic principles.

1.11 Which of the following phenomena have fuzzy ingredients?

(a) Flood traces on wadi sides

(b) Rainfall occurrence

(c) Inundation area

(d) Runoff coefficient

(e) Semiconfined aquifer

1.12 Are all the following statements fuzzy?

(a) IF rainfall occurs, THEN runoff starts.

(b) IF one digs 30 meters, THEN the groundwater table will be reached.

(c) IF rainfall intensity is high, THEN there is a possibility of groundwater recharge.

(d) IF sun shines, THEN there is evaporation.

(e) IF temperature is high, THEN agriculture will not be sufficient.

2 Linguistic Variables and Logic

2.1 GENERAL

Fuzzy logic has the ability to express the amount of ambiguity in human thinking and subjectivity, including natural language, and hence words, adjectives, and sentences in a comparatively undistorted manner. Linguistic variables are the most fundamental elements in human knowledge exposition and dissemination. The introduction of these variables provides an opportunity to formulate vague descriptions in natural language in a precise mathematical manner. Fuzzy concepts have a linguistic basis, and therefore the fundamental elements are words and their derivatives with adjectives. In some publications, the fuzzy operations are called *word computation* (Zadeh, 1999; 2001). It might seem rather strange to many readers but some words in daily life imply models, numerical values, ranges, and percentages. In this chapter such words in the hydrology domain are explained; their representation by formal and fuzzy sets are elaborated with examples.

Human rational thoughts become crystallized in scientific sentences, which include meaningful words implying events or objects. Humans transfer their opinions, hypotheses, theories, laws, and rules by words and sentences. Human behavior is formed and given certain direction by words and sentences. The treasure of information and data are the words and sentences, which must be understood properly for success in any task. Many people might think that except for scientists and engineers, other careers (including, among others, medicine, law, theology, literature, etc.) base career development on words and sentences, and in their writings there are no equations or numerical calculations. This is the reason why most of the time they are regarded as artists but not engineers or scientists. It is true that engineers base their opinions and conclusions on formulations and numerical information, but they also have at the back of their minds linguistic expressions that may enable them to work as artists, if they want to do so. This can be done using FL (fuzzy logic) principles.

This chapter explains the fuzzy content of the words, adjectives, and sentences on which the human thinking stages are based. Furthermore, approximate reasoning fundamentals are also explained with different scales of objects.

2.2 WORDS

Each physical or metaphysical object perception is expressed by a word, which may have more than one spelling, and in general, may have some scale. In this manner, each word reminds human beings of a collection (category) of similar objects with some personal differences (fuzziness) to a certain extent. For instance, the word *fluid* includes, by definition, any water that can take the shape of the reservoir it is put in, and also moves according to reservoir geometry and slope. The word *fluid* does not give any impression about numbers but provides crisp categorization of many objects into "fluid" and "non-fluid," where any object that is not water is "non-fluid." This is a two-way categorization of objects and it is referred to as classical logic (CL) categorization, which owes its systematic establishment to Aristotle with the Law of the Excluded Middle. Hence, "everything is known by its opposite" is a valid statement. However, knowing the properties of a fluid, many objects can be easily categorized as non-fluid; but according to the same definition, many fluids cannot be distinguished as water, or at least there are queries about their categorization as fluids. Such a categorization is mostly by visualization where the basic property is perceived through observations only. For instance, hydrochloric acid is also water in appearance but its close inspection gives the impression that it is a special type of water with its smell or the effect on the human mind and health. Hence, humans start to sub-categorize the water into refined categories such that there may also be sub-sub-categories. In this manner, it is possible to go from a general word to finer categories. This is similar to granulation of a mixed soil including "gravel," "pebble," "sand," "clay," and "silt." Each word as a name may have a different granulation for describing the same object in different scales.

A universe of discourse is made up of sets. For example, the universe of rock consists of different sets (sub-classes) such as "light rock," hard rock," "fractured rock," "sedimentary rock," "volcanic rock," "fissured rock," etc. Every stone is an element in the universe of rock.

On the other hand, adverbs are used in English in front of a word so as to specify its characteristics concerning the taste, appearance, weight, shape, size, quantity, etc. Not every two words specify the feature of the object such as red-river, salt-water, high-flood, and the like. It is necessary to be cautious in the use of adjectives that do not have much to say in the first place such as "dense," "intense," "flash," "high," "dry," "wet," etc. Adverbs often tell when, where, why, or under what conditions something happens or happened. They frequently end in -ly; however, many words and phrases not ending in -ly serve an adverbial function and an -ly ending is not a guarantee that a word is an adverb. The words *extremely, lengthy, monthly, severely,* and *neighborly* are adjectives. Adverbs can modify adjectives but an adjective cannot modify an adverb. Thus, one can say that "the hydrologists showed a really wonderful performance" and that "the variables showed a wonderfully casual attitude." Like adjectives, adverbs can have comparative and superlative forms to show degree:

1. If surface water flows "faster," then there is a danger of flood.
2. At the dam crest, the water in the dam is at its "highest" level.

It is important to notice that words such as *faster* and *higher* imply fuzziness. One often uses "more" and "most," "less" and "least" to show degree with adverbs such as in the following sentences, again with fuzzy contents:

1. Infiltration takes place "more quickly" at permeable surfaces.
2. The watersheds are the "most actively" arranged pieces of land for water management.
3. The hydrological calculations are regarded "less confidently" after extreme droughts.

Furthermore, the "as . . . as" construction can be used to establish adverbs that express sameness or equality:

1. The groundwater in the aquifers moves "as fast as" in the adjacent aquifer.
2. Evaporation rate is "as high as" the infiltration rate.

On the other hand, there are a handful of adverbs that do not end in -ly. In certain cases, the two forms have different meanings:

1. The flood wave arrived "late."
2. "Lately," hydrological modeling seems to be "more accurate."

In most cases, however, the form without the -ly ending should be reserved for casual situations:

1. Desertification is a "slow" phenomenon.
2. The average areal rainfall calculation is "wrong" compared with the measurements.
3. The flood arrived "sudden" and in "huge" volume to control cross-section.

Adverbs often function as intensifiers, conveying a greater or lesser emphasis on something. Intensifiers are said to have three different functions. They can emphasize, amplify, or tone down the meanings in the context of hydrological problem solving. They are related to fuzzy hedges, which are explained in Chapter 3.

Our daily lives pass with the spelling out of hundreds of words, many of which imply numerical values. Human beings became so automatic in daily linguistic activities that unconsciously they use words without fine and detailed thought filtering. For instance, if one says that "today is hot," he or she means some hidden numbers within a range that does not have clear boundaries. For someone in the Arabian desert, "warm" may imply around 30°C to 35°C, whereas in mid-latitudes it may mean 20°C to 25°C. The variable that represents "warm" in this case is the air temperature. When a word subsumes numbers, a mathematical framework starts to appear from its descriptive content. If the main word is *temperature*, it constitutes the universal set full of many sub-words (subsets), which can be broken down by additional words such as *cold, cool, moderate, warm, hot, very hot, extremely hot*, etc. Hence, words may be regarded as linguistic variables, which also imply some numbers. For such words, there is no formal framework to formulate it in the classical mathematical sense because *temperature* as a word is not a genuine numerical variable but rather a linguistic variable. If we can formulate the words in arithmetical or mathematical terms, then expert views can be treated for meaningful models.

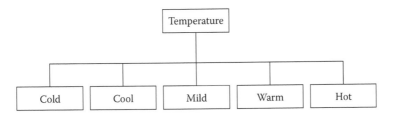

FIGURE 2.1 Word and adjectives.

In the above-mentioned example, the basic word *temperature* may be divided into comparatively less extensive sub-meanings (sub-words, subsets) as shown in Figure 2.1. Each these words is an adjective that describes the word *temperature*.

In Figure 2.1, the word *temperature* has specific attachments in different contexts. For instance, in set theory, "temperature" represents the universal discourse or universal set. In mathematics, it corresponds to a variable, which may be shown symbolically by T. In either case, namely as a universal set or mathematical variable, the word *temperature* reveals numerical implications, including even negative values. For instance, if it indicates air "temperature" for human survival, then the numerical values may not be "very small" or "big". Practically, one can claim that it might assume a maximum around 60°C and minimum of about −40°C. Herein, the words *around* and *about* mean approximation, and even *approximation* implies verbal and corresponding numerical uncertainty. The verbal uncertainty is fuzzy specification, and therefore the maximum or minimum temperature value is not known exactly but approximately, and may even change from individual to individual and from one location to another. In graphical representation, the "temperature" is the name of an axis along which there is a background number system as in Figure 2.2, including a zero value as a reference point. The ordered (ratio scale) numbers increase from the left toward the right. In the same figure, adjective words are also shown. The reader must notice that the adjectives are given in their natural sequence. Surely, the logical order of the adjectives has not an alphabetical but a hidden numerical foundation. If the adjectives are ordered alphabetically, then the final result appears as in Figure 2.3.

A comparison of Figures 2.2 and 2.3 shows that the former is plausible. The logical foundation behind this acceptance is the numerical implication of adjectives. It

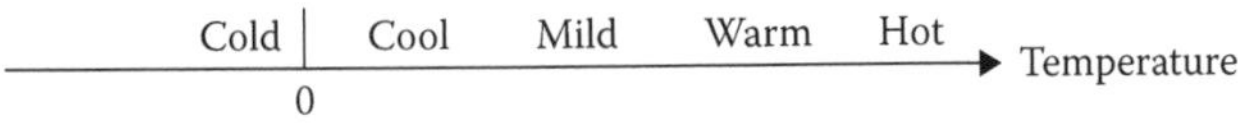

Cold | Cool   Mild   Warm   Hot → Temperature
0

FIGURE 2.2 Word and adjective variables.

Cold | Cool   Hot   Mild   Warm → Temperature
0

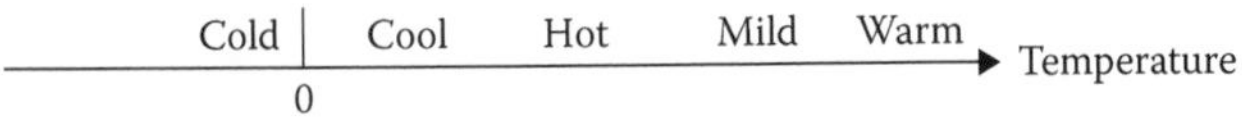

FIGURE 2.3 Word and adjective variables.

is possible to increase the number of adjectives as much as possible by considering different prefix words, such as *rather cold*, *very cold*, *very very cold*, etc. This means that each linguistic variable may have a set of adjectives, each of which can be refined as much as desired. An increase in the number of adjective words causes refinement (granulation) in the main word, that is, the variable domain description. However, the number of meaningful adjectives in the human sense is finite but computers may grasp even more meaningless adjectives such as *very very very cold*, *rather very extremely hot*, and so on. These meaningful adjective sequences for computers are explained in Chapter 3 with hedges.

Because the numerical implications of each word and its adjectives are not crisp (deterministic, certain), they are regarded as vague, incomplete, uncertain, which are collectively expressed by the word *fuzzy* and, accordingly, they are fuzzy words. By definition, any word that implies numerical values is a fuzzy word, which is also referred to as a *linguistic variable*. According to this definition, all the words in a language can be divided into two broad categories: as fuzzy words or non-fuzzy words. Examples of non-fuzzy words are water, hydrology, turkey, lion, confusion, spirit, lake, sea, etc. Fuzzy word examples include discharge, rainfall, evaporation, force, porosity, mass, acceleration, runoff, permeability, etc. In any text, identification of fuzzy words is possible by considering two features imbedded in them:

1. Fuzzy words imply direct numerical values, which are called in this case *numerical fuzzy words*. Examples include age, weight, distance, time etc.,
2. Some fuzzy words may be categorized verbally, and these are known as *categorical fuzzy words*. These fuzzy words also eventually subsume numerical values, which cannot be appreciated directly by everyone but they are defined according to specialization. Some examples are porosity, unit hydrograph, storativity, and specific drawdown, all of which have basic definitions. For instance, porosity is the ratio of the void volume in a material to the total volume, and therefore its values are confined between 0 and 1, inclusively, in a gray manner.

Example 2.1

The following is a paragraph that explains the rainfall phenomenon. Find the numerical and categorical fuzzy words and adjectives, if any, in this text:

The best way to obtain the hydrograph at any cross-section along a channel is to measure the runoff discharge by time, if possible, after an individual storm rainfall. The storm rainfall fingerprint is measured through a recording rain gauge, which gives the change of total rainfall height per unit time. It is then converted into the corresponding hyetograph. The watershed transforms such a hyetograph into a hydrograph, depending on the hydrological and geological features of the drainage basin.

Numerical fuzzy words are *discharge,* and *rainfall* because each one implies numerical values as mathematical variables or universal sets of all the discharges and rainfall amounts all over the world, respectively.

Among the categorical fuzzy words are *way* because it may be *best, worst, medium*, etc; *channel* because it may be categorized as *river, stream, creek*, or *rivulet*; *time* because it may have categories such as *early time, moderate time*, or *late time*; and *height* because it may assume categories such as *small, average, tall, big*, or *huge*.

Example 2.2

Assessment of groundwater resources in any area is not possible unless the geological water containing layers and their hydrogeological properties and parameters are not identified. Although groundwater hydrologists most often make groundwater resources evaluation, their approaches are based on the porous medium conceptualization of the geological formations. Porous medium enclaves spatial extension of a geological formation provided that it is composed of fine-, medium-, and coarse grains. Among these pieces are the voids, which might be filled by water, and these zones are referred to as *saturation zones*, which constitute the major reservoir of groundwater storage. The same situation is also valid for petroleum. The majority of the world"s groundwater reserves are within the sedimentary geological layers, and therefore all the early studies of groundwater occurrence, distribution, and movement were based on the porous medium conceptualizations. Chapter 3 provides a detailed explanation of porous medium from the groundwater existence of views.

Although there are no explicit numerical values in the above text, one can identify fuzzy words that imply numerical values. Among the numerical fuzzy words are *parameters, fine-grained, medium-grained*, and *coarse-grained*. On the other hand, *rock pieces, void, saturation*, and *porous medium* are the categorical fuzzy numbers.

Example 2.3

Identify the fuzzy words in the following hydrological text and make variable (universal set) and adjective (hedges) partition graphics similar to Figure 2.1:

Groundwater is forced to flow if hydraulic gradient exists throughout a geological formation. The groundwater flow velocity becomes of interest where the aquifer is endangered from any kind of pollution and also as a requirement of nuclear safety regulations to foresee the future behavior of the contaminants in the groundwater regime if a nuclear power plant accident occurs as it happened with the Chernobyl power plant in Russia. Physically, the velocity is defined as a ratio of distance to time, but in the case of groundwater it is almost impossible to trace the flow lines" actual path length as the water in a geologic formation moves in a random manner because of the tortoise of the medium.

The fuzzy words *force, gradient, velocity, distance* and *time* give the feeling that they have numerical values. On the other hand, categorical fuzzy words are *safety, contaminant, regime, tortoise*, and *medium*. It is not possible to find any adjective in this paragraph. The partition graphic of the fuzzy word *safety* is presented with logical sequence in Figure 2.4.

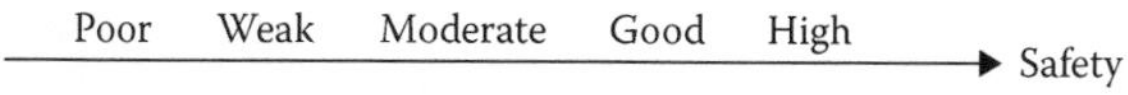

Poor   Weak   Moderate   Good   High
$\xrightarrow{\hspace{8cm}}$ Safety

FIGURE 2.4 Safety and its adjectives.

If the "safety" (S) is defined as a percentage of occurrence (probability, or relative frequency in the classical mathematics), then "poor," "weak," "moderate," "good," and "high" specifications may correspond to percentages such as $S < 0.25$, $0.25 < S < 0.50$, $0.50 < S < 0.75$, $0.75 < S < 0.90$, and $0.90 < S < 1.00$, respectively. In this manner, the fuzzy word *safety* is expressed categorically and numerically, and therefore it is now ready to be used in any arithmetical calculation or fuzzy modeling.

One very crucial point in this percentage partition of "safety" is that it is a crisp categorization because, as long as the "safety" is, say "good," the "safety" value is equally valid without distinction in the $0.75 < S < 0.90$ range. This seems rather illogical because there remains the question of whether 0.75 or 0.74 is completely "moderate" or completely "good". Hence, such crisp categorizations have problems at the transitional boundary between two adjacent categories. It is possible to conclude that although the fuzzy words are identified, they are still crisp in their numerical value attachments. The question that remains is how to make these fuzzy words uncrisp; that is, how to fuzzify their numerical values so that even within the same range (category) there will be preference between the numerical values. This is the fuzzification problem of the fuzzy words, which is achieved by means of membership functions (MFs) and is explained in Chapter 3.

Although fuzzy word content is uncertain, it can be represented either crisply or fuzzily. In the former case, the numerical content of a fuzzy word does not make any distinction between the values, whereas the latter case is based on the distinction (preference, inequality) between the content values. In all the classical studies of modeling in hydrology, the former alternative is adopted through the set theory and mathematical principles. However, in this book, the fuzzy sets are of prime importance for representing the content of fuzzy words.

2.3 LINGUISTIC VARIABLES

Natural languages are the most powerful means of information communication among humans, but such information is vague and imprecise. Despite this vagueness in natural languages, users of the same language have little trouble in understanding each other. For instance, let us adopt "temperature" as a linguistic variable and think about the meaning of "high temperature." Different individuals might imagine different numerical Celsius degrees such as 35°C, 37.5°C, 45°C, etc., but all remain under the linguistic umbrella of "high temperature" with different degrees of experience. Hence, computers cannot understand this umbrella information without a number range and presently require crisp numerical or interval values without any difference between the elements, such as 40°C, or from 35°C to 45°C, respectively. In the first case, the umbrella is not open and is very crisp; but in the second case, the umbrella is open with a collection of several elements.

During actual rainfall, it does not mean that each individual under the umbrella will not get equally wet. The ones right in the middle of the umbrella will not get wet; however, as individuals stand away from the center, their degrees of "wetness" will different. It is possible to deduce from this argument that vague statements have a set of values that are within an interval but with different degrees of "belongingness" to this interval. This results in a consensus between individuals with differing perceptions of the concept of degree, by placing them under an overarching umbrella concept. An example would be "high temperature," where each person agrees with the overall idea, but relates it to a different Calsius degree value. Hence, regardless of their "temperature" variability, individuals do not require identical definitions of the term *high temperature* to communicate effectively. The power of fuzzy principles lies in their ability to use linguistic variables, rather than quantitative variables, to represent imprecise concepts. The intuition and judgment of any researcher about the phenomenon concerned play a major role in modeling, and hence the fuzzy concepts as fuzzy sets and models provide digestion of imprecise, vague, and ambiguous information.

2.4   SCIENTIFIC SENTENCES

Conscious direction of attention toward an external object causes the object to be received by the mind into the realm of our fuzziness, which causes, in sequence, perception, experience, feeling, thinking, understanding, knowing, and finally acting for a meaningful description and analytical solution. It was stated by Zadeh (1971) that fuzzy statements are the only bearers of meaning and relevance.

Words and sentences constitute a language under a set of strict grammar rules for meaningful communication between the common users of the language. Categorization is a common human conceptualization. The naming and description of the categorizations may have different words but the philosophical and logical concepts remain almost the same in any language. The human mind cannot solve any problem numerically if the linguistic solution structure is not well understood. This is the general rule in scientific research. Unfortunately, many people think that knowing higher-level mathematics will help solve the problems in a better way. In fact, this was the major difficulty and complexity that Zadeh (1965), the founder of FL, encountered during his modeling studies. He realized that the addition of any extra mathematical equation with refined and better solution expectations led to unmanageable mathematical giants that did not provide significant additional benefit. Then he thought that there must be a simple, logical way to solve the problem linguistically and he coined the term *fuzzy logic* (FL). The CL of blacks and whites only (mathematical equations) may lead to unmanageable forms. Even the analytical solutions of any natural or artificial event require a set of assumptions and idealization to reduce its complexity to the perception level of human beings who can then solve the problem approximately with the use of mathematical formulations in a crisp manner.

Sentences provide additional information about the nouns and words (categories) concerning their spatial or temporal positions, states, and interrelations. The sequence of words in a meaningful manner makes the sentences but there are certain native language grammar rules according to which the sentences take shape. In English, the simplest form of the sentence has three components: subject (static) +

verb (dynamic) + object (description). The first word in the sentence is a noun, that is, a category. The last word is the description of this name. For instance, the sentence

"Flood is intensive"

is a proper sentence in which two words are joined by an intermediate word that does not have any physical implication in this case. It is also possible to express the same idea as

"Intensive flood"

which is time independent and more general than the previous one. Both have a fuzzy meaning. The latter does not include a verb and hence cannot be labeled as a proper sentence. In English, only "verb sentences" are available; but in some other languages, such as in Arabic, there are also "name sentences." Sentences provide extra information for *word computation* or *soft computation*. In the given example, the intensity of flood has not been sub-categorized, and therefore it has vagueness, ambiguity, and a broad uncertainty. The reader is advised to compare the following sentences with the original one:

- "Flood is moderately intensive."
- "Flood is very intensive."
- "Flood is almost intensive."
- "Flood is rather intensive."
- "Flood is highly intensive."
- "Flood is approximately intensive."
- etc.

The first impression one gets when considering these sentences simultaneously is that there are differences in the intensities of the flood. The reader may try to order these sentences in increasing magnitude depending on his or her linguistic and expert backgrounds. It is not possible to obtain absolutely the same sequence from each reader. This also indicates that sentences include uncertainty, that is, fuzziness. Examples can be increased and the reader may feel the meaning of *word computation* without numbers but perceptions. This is the very basis of the FL modeling in this book. Of course, because computers understand through numbers only, these perception computations by words or sentences must be converted into numbers so that humans and computers can understand each other. Such matches between human and computer understandings are provided by the MFs, which are discussed in Chapter 3.

2.5 FUZZY SCALES

An approximate quantity can be attached to any object by perceptions, which are expressed in words and reflect the magnitude or amount of some characteristic. So, each word has an appreciable and attachable but vague relative quantity for describing the events or objects. For instance, a silt particle of sand size is not plausible but it can be as big as a clay particle. Hence, the human mind does not accept sand-size silt, due to one's conscious experience perhaps based on crisp logic. In classical studies, the manner in which numerical values are assigned determines the scale of measurement and it is necessary to have an instrument for exact measurements. However, in the fuzzy context, precise measurements are not very meaningful and, more often, approximate values should be appreciated prior to the solution of any problem. Hence, perception and appreciation of various scales become significant, say for instance, for the words *gravel*, *pebble*, *sand*, *clay*, and *silt*—hence different scales.

In general, there are four scales that provide the quantitative values with qualitative appreciation: the nominal, ordinal, interval, and ratio scales. Each contains more information content than its predecessor. Each observation provides the list of objects and corresponding words in the observer's native language. Quantitative appreciations about the objects are also gained during the observation process.

2.5.1 NOMINAL SCALE

This is a scale that is appreciable by a noun or name. It consists of classifying observations into mutually exclusive categories of equal rank (belongingness degree, membership degree, or MD). These categories can be identified by nouns, such as "gravel," "pebble," "sand," "clay," and "silt," and symbolized by the letters "G," "P," "S," "C," and "s," respectively. One can also use geometrical symbols, as in Figure 2.5.

Herein there is no connotation that the first one is twice as large as the second, and so on. According to CL, each one of these nouns can be dichotomized into two mutually exclusive and exhaustive sets as "gravel" and "not-gravel," "pebble" and "not-pebble," "sand" and "not-sand," "clay" and "not-clay," and "silt" and "not-silt". In reality, neighboring pieces are not mutually exclusive because gravel cannot be distinguished crisply from pebbles, and therefore there is overlap between two categories. This point implies that nominal scales are fuzzy in character.

Nominal variables allow for only qualitative classification. That is, they can be measured only in terms of whether the individual items belong to some distinctively different categories, but one cannot quantify or even rank the objects in each category. For example, all one can say is that two drainage basins are different in terms of, say, climate variable, but one cannot say without observation which one "has a favorable" climate, say, for agriculture. Typical examples of nominal variables are climate, morphology, vegetation, land use, etc.

In classical statistics, a frequency distribution shows the number of observations in each of the several classes (ranges of values), and it is usually used to analyze data measured on a nominal scale. A variable is any measured characteristic or attribute

FIGURE 2.5 Different earth pieces.

TABLE 2.1
Köppen Climate Classification System

Climate Type (Nominal scale)	Description
A	Tropical moist climates: all months have average temperatures above 18°C
B	Dry climates: with deficient precipitation during most of the year
C	Moist mid-latitude climates: with mild winters
D	Moist mid-latitude climates: with cold winters
E	Polar climates: with extremely cold winters and summers

that differs for different subjects. For example, if the sizes of 30 catchments are measured, then the size would be a variable, which is measured on a nominal scale and often referred to as a *categorical* or *qualitative variable*. These variables are vague and imprecise.

In some cases, linguistic information is converted into nominal scales for the description of the phenomenon concerned, even by non-experts. For instance, the Köppen climate classification system is the most widely used system for classifying the world"s climates. Its categories are based on the annual and monthly averages of temperature and precipitation. The Köppen system recognizes five major climatic types, wherein each type is designated by a capital letter as in Table 2.1.

2.5.2 Ordinal Scale

A set of data is said to be ordinal if the values (observations) can be ranked (put in order) or have a rating scale attached. One can count and order, but not measure ordinal data.

The classification of runoff into different types ("wet," "moderate," and "dry") is an example of nominal scaling. The number of observations in each state of a nominal system can be counted. For instance, there are two possibilities that each year will either be either "wet" (W) or "dry" (D). If 3 years are considered, then the nominal sequences provide the outcome of three consecutive observations in $2^3 =$ 8 mutually exclusive alternatives as WWW, WWD, WDW, DWW, WDD, DWD, DDW, or DDD. It is therefore possible to count or calculate the number of possible events with the nominal scales. For example, the objective of the Palmer Drought Severity Index (PDSI) was based on the ordinal scale and provided measurements of moisture conditions that were standardized so that comparisons using the index could be made between locations and months (Palmer, 1965). This is a meteorological drought index and it responds to weather conditions that have been "abnormally dry" or "abnormally wet". When conditions change from dry to normal or wet, for example, the drought measured by the PDSI ends without taking into account the stream flow, lake and reservoir levels, and other longer-term hydrologic impacts (Karl and Knight, 1985). The PDSI consists of 11 descriptions arranged in order from 1 to 11, as in Table 2.2.

TABLE 2.2
Palmer Classification

Groups	Palmer Value	Description
1	4.0 or more	"extremely wet"
2	3.0 to 3.99	"very wet"
3	2.0 to 2.99	"moderately wet"
4	1.0 to 1.99	"slightly wet"
5	0.5 to 0.99	"incipient wet spell"
6	0.49 to −0.49	"near normal"
7	−0.5 to −0.99	"incipient dry spell"
8	−1.0 to −1.99	"mild drought"
9	−2.0 to −2.99	"moderate drought"
10	−3.0 to −3.99	"severe drought"
11	−4.0 or less	"extreme drought"

Source: Palmer, W.C. 1965. *Meteorological Drought*. Research Paper No. 45, U.S. Department of Commerce Weather Bureau.

The categories for an ordinal set of data have a natural order. For example, suppose a group of hydrologists is asked to rate varieties of droughts and classify each drought according to descriptions given in Table 2.2. A rating of 7 indicates more severity and risk than a rating of 6, so such data are ordinal. However, the distinction between neighboring points on the scale is not necessarily always the same. For instance, the difference in severity expressed by a rating of 2 rather than 1 might be much less than the difference expressed by a rating of 4 rather than 3. There is no "true" zero point for ordinal scales because the zero point is chosen arbitrarily.

On the other hand, the Standardized Precipitation Index (SPI) is designed to express explicitly the fact that it is possible to simultaneously experience wet conditions on one or more time scales and dry conditions at other time scales, often a difficult concept to convey in simple terms to decision makers (McKee et al., 1993; 1995). Consequently, a separate SPI value is calculated for a selection of time scales, covering the last 1, 2, 3, 4, 5, 6, 7, 8, 9, 10, 11, 12, 15, 18, 24, 30, 36, 48, 60, and 72 months and ending on the last day of the latest month. The description of drought based on SPI is presented in Table 2.3.

Ordinal variables allow one to order the items that he or she measures in terms of which has less and which has more of the quality represented by the variable, but still they do not allow one to say "how much more." One can say that nominal measurement provides less information content than ordinal measurement, but one cannot say "how much less" or "how this difference compares" to the difference between ordinal and interval scales.

In ordinal scales, each description is expressed by fuzzy words, which include uncertainty in terms of vagueness, and each description can be represented by a fuzzy set as explained in Chapter 3.

TABLE 2.3
SPI: Standardized Precipitation Index

Classes	SPI Values	Description
1	>2.00	"extremely wet"
2	1.5 to 1.99	"very wet"
3	1.0 to 1.49	"moderately wet"
4	−0.99 to 0.99	"near normal"
5	−1.0 to −1.49	"moderately dry"
6	−1.5 to −1.99	"severely dry"
7	< −2.00	"extremely dry"

Source: McKee, T.B., N.J. Doesken, and J. Kleist 1995. Drought Monitoring with Multiple Time Scales, *9th Conference on Applied Climatology*, pp. 233–236.

2.5.3 INTERVAL SCALE

The interval scale is a scale of measurement where the distance between any two adjacent units of measurement (or "intervals") is the same but the zero point is arbitrary. Scores on an interval scale can be added and subtracted but cannot be meaningfully multiplied or divided. For example, the time interval between the start of any water year in the last decade is the same and the zero point is arbitrary. Other examples of interval scales include the heights of tides and the measurement of longitude. PDSI and SPI scales are fuzzy, and change from one interval to another. In the interval scale, the intervals between two successive granulations are constant. In this scale, there is not a natural zero where the magnitude is nonexistent. For instance, temperature degradation has similar intervals between 230°C to 231°C and 1°C to 2°C. The zero temperature also represents a natural event of coldness. Hence, zero is attached to a natural event of water freezing in the Celsius interval scale, whereas in the Fahrenheit degradation, the zero is attributed to the temperature of a snow and salt mixture. This indicates that any interval scale is relative to a natural event.

Each "higher" level of measurement includes the measurement principle of the "lower" level of measurement. For example, the numbers 8 and 9 in an interval scale indicate that the object assigned a 9 has more of the attribute being measured than does the object assigned an 8 (ordinal property), and that all persons assigned a 9 have equivalent amounts of the attribute being measured (nominal property). This also implies that one can do lower-level statistics on higher-level measurement scales.

2.5.4 RATIO SCALE

The ratio scale has a true zero point and also equal intervals. Any physical quantity with a zero ratio scale does not exist. The ratio scale may have different units and it is possible to convert from one unit to another with a conversion factor. These are the scales where all the mathematical, probabilistic, statistical, and stochastic calculations in hydrology are performed with CL. Interval and ratio scales may be used interchangeably.

TABLE 2.4
The Measurement Principles

Nominal	Ordinal	Interval	Ratio
People or objects with the same scale value are the same on some attribute. The values of the scale have no "numeric" meaning in the way that one usually thinks about numbers.	People or objects with a higher scale value have more of some attribute. The intervals between adjacent scale values are indeterminate. Scale assignment is by the property of "greater than," "equal to," or "less than."	Intervals between adjacent scale values are equal with respect to the attribute being measured. For example, the difference between 8 and 9 is the same as the difference between 1976 and 1977.	There is a rational zero point for the scale. Ratios are equivalent, in that the ratio of 2 to 1 is the same as the ratio of 36 to 18.
FUZZY	**FUZZY**	**CRISP**	**CRISP**

Typical examples of ratio scales are measures of time or space. For example, as the flood discharge scale is a ratio scale, not only can one say that a flood of 250 m³/sec is higher than one of 125 m³/sec; but one can also correctly state that it is twice as high. Interval scales do not have the ratio property. Most statistical data analysis procedures do not distinguish between the interval and ratio properties of the measurement scales.

Any object in the hydrological sciences can be measured by an instrument, and the results appear in numbers. Table 2.4 shows the measurement principles concerning each scale. The crisp or fuzzy characteristic of each scale is also indicated in the same table.

Stevens (1951) classified not just simple operations, but also statistical procedures according to the scales for which they were "permissible". A scale that preserves meaning under some class of transformations should be restricted to statistics whose meaning would not change were any of those transformations applied to the data. By this reasoning, analyses on nominal data, for example, should be limited to summary statistics such as the number of cases, the mode, and contingency correlation, which require only that the identity of the values be preserved. Permissible statistics for ordinal scales included these plus the median, percentiles, and ordinal correlations—that is, statistics whose meanings are preserved when monotone transformations are applied to the data. Interval data also allowed means, standard deviations (although not all common statistics computed with standard deviations), and product moment correlations because the interpretations of these statistics are unchanged when linear transformations are applied to the data. Finally, ratio data allow all of these plus geometric means and coefficients of variation, which are unchanged by rescaling the data. In summarizing this argument, Luce (1959, p. 84) said:

> … the scale type places [limitations] upon the statistics one may sensibly employ. If the interpretation of a particular statistic or statistical test is altered when admissible scale transformations are applied, then our substantive conclusions will depend on which

arbitrary representation we have used in making our calculations. Most scientists, when they understand the problem, feel that they should shun such statistics and rely only upon those that exhibit the appropriate invariance for the scale type at hand. Both the geometric and the arithmetic means are legitimate in this sense for ratio scales (unit arbitrary), only the latter is legitimate for interval scales (unit and zero arbitrary), and neither for ordinal scales.

2.6 FUZZY LOGIC THINKING STAGES

Any scientific thinking has three major steps: namely, imagination, visualization, and idea generation. Figure 2.6 indicates the steps necessary in a complete thinking process (Şen, 2008). Each one of the steps cannot be explained in a crisp manner and each individual, depending on his or her capabilities, may benefit from this sequence.

The imagination step includes setting up suitable hypotheses for the problem at hand and the purpose of the visualization step is to defend the representative hypotheses. Scientists typically use a variety of representations, including different kinds of figures (geometry), to represent and defend the hypotheses. Scientific hypothesis justification is possible only through the understanding of visual representation and, if necessary, modification of the hypothesis should be in progress. On the basis of hypothesis, the scientists behave as philosophers by generating relevant ideas and their subsequent dissemination, which should include new and even controversial ideas, so that other scientists can surpass and further elaborate on the basic hypotheses. Whatever the means of thinking are, the scientific arguments are expressed by linguistic expressions prior to any symbolic and mathematical abstractions. In particular, in hydrological sciences, the visualization stage is represented by algorithms, graphs, diagrams, charts, and figures, which include a tremendous amount of condensed linguistic information.

Scientific visualizations have been conducted with geometry since the very beginning of scientific thought. This is the reason why geometry was developed and recognized by early philosophers and scientists over any other scientific tool (such as algebra, trigonometry, or mathematical symbolism). Al-Khawarizmi (died 840 A.D.), who is known in the West by his Latinized name "algorithm," solved second-order equations by considering geometric shapes (Şen, 2006). For example, he visualized x^2 as a square with side equal to x, and any term such as ax is considered

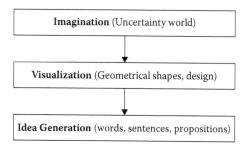

FIGURE 2.6 Thinking gradients.

a rectangle with base length x and height equal to a. This geometrical thinking and visualization made him the father of "algebra." All his discussions were explained linguistically.

All the conceptual models deal with parts of something that is perceived by the human mind. Of course, among the meaningful fragments of the phenomenon, there exist clear or hidden interrelationships, which are there for the exploration of the human intellectual mind. Such possible relationships can be explained by a set of fuzzy rule statements (see Chapter 6). Fragments of thinking, sensations, thoughts, and perceptions serve collectively to provide partial and distorted conceptual models of reality in representing a perceived human-mind-produced world.

Although human wonder and mind are the sources of fuzziness, they also serve to overcome problems with human experience, which can be regarded as expert views. The fuzzy concepts in understanding complex problems depend on observations, experiences, and conscious expert views. When problems are solved, there is always fuzziness related to them that paves the way for future development. Thus, scientific solutions cannot be taken as absolute truths in a positivistic manner.

Precise knowledge is possible only when a phenomenon or process is isolated from its sarroundings, again with a set of restrictive assumptions that render the problem into the world of certainty by ignoring all fuzzy features. For instance, the runoff coefficient is a multiplier applied to the deterministically (CL) calculated peak discharge according to rational formulation in hydrology. Thus, by effectively "over-engineering" or "under-engineering" the design by strengthening components or including redundant systems, a runoff coefficient accounts for imperfections in hydrologic calculations, flaws in assembly, geomorphologic and geologic degradation, and uncertainty in discharge estimates. In fact, the runoff coefficient includes an "ignorance component" due to the exclusion of all fuzzy information about the hydrologic design. However, FL and the system help solve the hydrologic design problem without considering the runoff coefficient because the solution is based on fuzzy, uncertain information (see Chapter 8).

There are no isolated phenomena and processes in hydrology, and any knowledge about them is always fuzzy. The significance of fuzziness opens ways for changes, evolution, growth, and continuous scientific development. Figure 2.7 documents the steps in fuzzy thinking and problem solving (Şen, 2007).

Fuzzy concepts concentrate on the study of the human mind possibilities to know external objects by collecting information through observations or readings. Once the collection of such fuzzy linguistic information is completed, then human inquiry can expand the field of understanding in different directions and, in the mind, the objects and their different visible properties are expressed first by words. Each item concerning the phenomenon under investigation is described by a word or a set of fuzzy words (sentences, statements, propositions). This is equivalent to the categorization of the objects into different classes, again in a fuzzy manner; and at this stage, CL cannot be helpful. For instance, when some objects are labeled by a word, say *river*, one is certain that there is fuzzy uncertainty in this labeling.

A fuzzy perception is an assessment of a physical condition that is not measured with precision, but assigned an intuitive value. It is asserted that everything in the

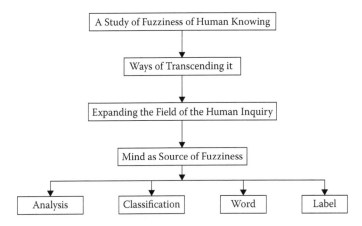

FIGURE 2.7 Fuzzy concepts.

universe has some fuzziness, no matter how good the measurement equipment is. By using meaningful words to name the fuzzy description, the construction of a hydrologic process is easy to understand and can be built up intuitively.

Humans are very good at recognizing by eye what they are looking at but computers are better at counting and measuring. FL is very helpful in guiding computers to find the right thing to measure and calculate. Real-world attributes are known by human perceptions through quality and quantity appreciations linguistically and/or by measurements. Different questions may be asked about individual or joint behaviors of these attributes. Humans continue to acquire knowledge by perception, which is a never-ending process. Fuzziness is a paramount characteristic of human perception that challenges humanity and propels the search for truth and understanding the secrets of reality. The fuzziness in human perception reveals ways of transcending it, and thus expanding the field of human inquiry.

Fuzzy impressions and concepts are generated by the human mind, and it divides the seeable hydrological, environmental, or engineering reality into fragments and categories, which are fundamental ingredients in the classification, analysis, and deduction of conclusions after labeling each fragment with a word such as "name," "noun," or "adjective". The initial labeling by words is without interrelation between various categories. These words have very little to do with the wholeness of reality. Hence, common linguistic words help us imagine the same or very similar objects in our minds in a fuzzy manner.

Every act of holistic understanding is inevitably fuzzy. Fuzziness and truth are not mutually exclusive, as is assumed in the CL, but they do go hand in hand in every aspect of scientific research.

When consciously directing one's attention toward an external object, the object enters into the realm of one's fuzziness of perception, which includes feeling, thinking, understanding, experiencing, knowing, and applying. The levels of fuzziness correspond to the levels of one's capacity for understanding and depend on the levels of consciousness.

Measured non-fuzzy data constitute one of the primary inputs for FL models. Examples are temperature measured by a thermometer, rainfall measured by rain gauges, groundwater levels measured by sounders, etc. Additionally, humans with their fuzzy perceptions could also provide input with linguistic statements.

Common sense dictates that some form of empiricism is essential to make sense of the world. In traditional quantitative educational training, classical dualism (as the tension between subjectivity and objectivity) is often addressed by adopting an objectivist, empiricist, or positivistic approach, and then by applying a scientific research design. Even based on CL, scientific thinking starts in an entirely subjective medium. Subjective thinking penetrates the objectivity domain by time through imagination and visualization, and hence there is not a crisp line between subjectivity and objectivity. Empirical works, which are based on either observations or measurements as experimental information, help decrease the degree of subjectivity to the benefit of objectivity degree. In a way, none of the scientific formulations obtained up to now are completely crisp but they are regarded as crisp information, provided that the fundamental assumptions (such as mutual exclusiveness and exhaustiveness) are taken into consideration. The crispness of any scientific information can be shacked by modifying one of the basic assumptions. This implies that all the scientific principles are not completely crisp, but include vagueness, incompleteness, and uncertainty—even to a small extent—and hence they can be considered fuzzy by nature or by human understanding.

In everyday life, human beings make many predictions, especially on the basis of qualitative data and past experience. Additionally, expert opinions help shape and refine such predictions in addition to the mutual discussion and confidence. In predictions there are similarities, which are the input information about the phenomenon concerned, the output clues, and the logical connectivity between these two sources of information. On the basis of certain cues, it is possible to make judgments about some target property, that is, output information. The default of these judgments is the commonly available scientific thinking and its sublime version of logic (CL or FL) leading to rational results. This provides the ability for any individual to develop actuarial models for various real-life prediction problems.

It is possible to make predictions either by CL mathematical models or FL expert views. A basic question is: Are the predictions of human experts more reliable than the predictions of mathematical models? Experts make their predictions on the basis of the same evidence as mathematical foundations, but additionally they consider the usefulness of the linguistic data in the form of vague statements in the adjustment of the final model. Such vague information cannot be digested by a CL mathematical model because any sort of uncertainty is defuzzified, that is, rendered into crisp numerical forms. It is, therefore, expected that fuzzy modeling by experts considering vague information is more successful than mathematical models, which are valid for ideal cases under the validity of a set of assumptions. Among the most important problems are natural phenomena predictions because they have the following properties:

1. Even the best mathematical models are not particularly reliable.
2. The best results seem to be reasonable predictions but somewhat unsafe, and therefore an interval of confidence is necessary.

Similar principles are also valid for geological, hydrological, meteorological, and atmospheric environmental phenomena. To move understanding toward a deeper and broader grasp of complexity, the emergent meanings must be neither stable nor unstable—that is, stable enough to rely upon them when generating hypotheses, concepts, and emotional attitudes, and unstable enough not to allow these concepts and attitudes to harden and become dogmas and addictions. In other words, after scientific thinking, meanings need to be fuzzy (flexible), ready to immediately respond to the changes continuously occurring in each of the countless dimensions of reality.

2.7 APPROXIMATE REASONING

Reasoning is the most important human brain operation that leads to creative ideas, methodologies, algorithms, and conclusions, in addition to a continuous process of research and development. The reasoning stage can be reached provided that there is stimulus for the initial driving of mental forces. Ignition of pondering on a phenomenon comes with the physical or mental effects that control an event of concern. These effects trust imaginings about the event and initial geometrical sketches of the imaginings by simple geometries or pieces and connections between them (Chapter 3). In this manner, the ideas begin to crystallize and they are conveyed linguistically by means of a native language to other individuals to get their criticisms, comments, suggestions, and support for the betterment of mental thinking and scientific achievement.

Approximate reasoning helps resurface information technology, where it provides decision support and expert systems with powerful reasoning bound by a minimum of rules, and it is the most obvious implementation for FL in the field of artificial intelligence. It was already explained how one can easily relate logic to ambiguous linguistics in the form of different fuzzy words such as *very*, *small*, *high*, and so on. Such flexibility allows for rapid advancements and easier implementation of projects in the field of natural language recognition. FL brings not only logic closer to natural language, but closer to human or natural reasoning. Many times, knowledge engineers must deal with very vague and common-sense descriptions of the reasoning leading to a desired solution. The power of approximate reasoning is to perform reasonable and meaningful operations on concepts that cannot be easily codified using a classical approach. Implementing FL will not only make the knowledge systems more user friendly, but it also will allow programs to justify better results.

Finally, all conclusions must be expressed in a language that can then be converted into universally used symbolic logic based on the principles of mathematics, statistics, or probability statements. This explanation shows that FL is followed by symbolic logic (mathematics). Unfortunately, in many educational systems all over the world, this sequence of language and symbolism is turned into the sequence of symbolism (mathematics) first and then linguistic understanding, which is against the natural reasoning abilities of humans. This is especially true for countries or societies that are trained with symbols. When one returns to their community, the first difficulty is to convey the scientific messages in his or her language, and therefore, in order to avoid such a dilemma, the teacher bases the explanation on symbolic logic. This is one of the main reasons why scientific thinking and reasoning are almost missing in many institutions all over the world. The avoidance of such a problem is approximate reasoning, where the facts are explained through natural languages first.

Subjectivity, that is, dependence on personal thoughts, is greatest at the perception stage; as one enters the visualization domain, the subjectivities decrease; and at the final stage because the ideas are exposed to other individuals, the objectivity becomes at least logical but still there remains some uncertainty (vagueness, incompleteness, missing information, etc.) and hence the final conclusion is not crisp but fuzzy. Fuzzy reasoning in water resources (hydrological and earth scientific domains) always exists, but in the classical and mechanical approaches it is deleted artificially by idealizations, isolations, simplifications, and assumptions.

The CL renders the final stage in solutions into crisp forms by defuzzification, which means neglecting all the uncertainties either through assumptions or through a safety factor or confidence interval in many engineering solutions. Crisp reasoning conclusions do not provide a soft domain for further research, especially in hydrology. Therefore, classical hydrological methodologies and formulations are fragile, hard, and their consequences are difficult to accept. To avoid the crispness, statistics and axiomatic probability concepts are suggested but they are also based on CL, where the consequences are black and white without gray tones, which are available in approximate reasoning through FL principles and modeling.

REFERENCES

Karl, T.R. and Knight, R.W. 1985. *Atlas of Monthly Palmer Hydrological Drought Indices (1931–1983) for the Contiguous United States.* Historical Climatology Series 3–7, National Climatic Data Center, Asheville, NC.

Luce, R.D. 1959. On the possible psychophysical laws, *Psycholog. Rev.* 66: 81–95.

McKee, T.B., Doesken, N.J., and Kleist, J. 1993. The Relationship of Drought Frequency and Duration to Time Scales, Preprints, *8th Conference on Applied Climatalogy,* Anaheim, CA, pp. 179–184.

McKee, T.B., Doesken, N.J., and Kleist, J. 1995. Drought Monitoring with Multiple Time Scales. Preprints, *9th Conference on Applied Climatology*, Dallas, TX, pp. 233–236.

Palmer, W.C. 1965. *Meteorological Drought.* Research Paper No. 45, U.S. Department of Commerce Weather Bureau, Washington, D.C.

Stevens, S.S. 1951. Mathematics, measurement, and psychophysics. In S.S. Stevens, Ed., *Handbook of Experimental Psychology.* New York: John Wiley & Sons.

Şen, Z. 2006. *Batmayan Guneslerimiz (Our Unsettable Suns).* Altın Burc Publications, 100 pp. (in Turkish).

Şen, Z. 2007. *Fuzziology-Fuzzy Philosophy of Science and Education. Critical Review.* A Publication of the Society for Mathematics of Uncertainty, Paul P. Wang, Ed., pp. 33–44.

Şen, Z. 2008. *Wadi Hydrology.* CRC Press, Boca Raton, FL, Taylor & Francis Group, 347 pp.

Zadeh, L.A. 1965. Fuzzy sets. *Informat. Control* 8: 338–353.

Zadeh, L.A. 1971. Towards a theory of fuzzy systems. In R.E. Kalman and N. DeClaris, Eds., *Aspects of Network and System Theory* New York: Holt, Rinehart, and Winston, pp 209–245.

Zadeh, L.A. 1999. From computing with numbers to computing with words—From manipulation of measurements to manipulation of perceptions. *IEEE Trans. Circuits and Syst.* 45(1): 105–119.

Zadeh, L.A. 2001. A new direction in AI-towards a computational theory of perceptions. *Am. Assoc. Artificial Intelligence Mag.* Spring 2001, pp. 73–84.

PROBLEMS

2.1 Identify the numerical and categorical fuzzy words in the following paragraph:

"Seeding of tropical cumulus clouds, and indeed any clouds, requires that they contain supercooled water, that is, liquid water colder than zero Celsius. Introduction of a substance, such as silver iodide, that has a crystalline structure similar to that of ice will induce freezing. In mid-latitude clouds, the usual seeding strategy has been based upon the vapor pressure being lower over water than over ice. When ice particles form in supercooled clouds, they grow at the expense of liquid droplets and become heavy enough to fall as rain from clouds that otherwise would produce almost none."

2.2 Find the categorical fuzzy words in the following paragraph:

"The process of desertification is a slowly creeping phenomenon, which takes place in any area during long time durations due to different reasons. The first indications are due to variations in the weather or meteorological parameters. In general, desertification implies decrease in some interesting meteorological and agricultural quantities such as rainfall amounts, vegetation coverage, surface water extensions, groundwater level drops, and crop yields. On the other hand, increases in the weather temperature, sand coverage, areal drought coverage, urban area expansion, and sedimentation amounts all imply desertification. In simple terms, the historical records of these variables either as time series measurements through local land surface instruments or satellite images should lead to increasing or decreasing trends depending on the quantity concerned and the monitoring network."

2.3 Identify the fuzzy words in the following paragraph:

"Initially, high water table areas must be examined. Most often, these areas are underlain by a water table mound; the status of hydrographs in bores around the mound would indicate its rate of spread and the long-term prognosis. Trends identified from borehole hydrographs are most reliable for long periods of records. In order to account for seasonal climatic fluctuations, the record period should extend over a number of years. In most of the arid regions, records are available most often for few years, and many are considerably shorter. Therefore, the assessment of trends should be considered a preliminary data treatment. Statistical trends analysis can be affected with the coming of additional, more available information."

2.4 Underline the fuzzy words in the following passage:

"There are no perennial surface flows in Saudi Arabia and the annual runoff volume is usually concentrated in the form of flash floods of short duration but sizable magnitude, which mostly occur during the expected rainy seasons in winter and spring. Because of ground and climate conditions, rainfall is immediately converted to runoff, causing floods or flash floods due to the following reasons.

1. In the upper parts of the catchments, where there is almost no soil to trap water, the slopes tend to be steep and the rocks are nearly impervious. Therefore, infiltration losses and retention by filling the depressions are minor.
2. In the foothills of catchments, the high intensity of the rain seals the surface of rather bare soil quickly; consequently, only a shallow depth of penetration of soil moisture is achieved before pounding and the onset of surface runoff."

2.5 Find the probabilistic words in the following paragraph:

"The best way to obtain the hydrograph at any cross section along a channel is to measure the runoff discharge by time, if possible, after an individual storm rainfall. The storm rainfall fingerprint is measured through a recording rain gauge, which gives the change in total rainfall height per unit of time. It is then converted into the corresponding hyetograph. The watershed transforms such a hyetograph into a hydrograph depending on the geological and hydrological features of the drainage basin."

2.6 The following is a paragraph that explains the rainfall phenomenon. Find the numerical and categorical fuzzy words and adjectives, if any, in the following text:

"The best way to obtain the hydrograph at any cross section along a channel is to measure the runoff discharge by time, if possible, after an individual storm rainfall. The storm rainfall fingerprint is measured through a recording rain gauge, which gives the change of total rainfall height per unit of time. It is then converted into the corresponding hyetograph. The watershed transforms such a hyetograph into a hydrograph, depending on the geological and hydrological features of the drainage basin.

2.7 Order the following words according to their numerical implications:
   (a) "cold"
   (b) "hot"
   (c) "warm"
   (d) "cool"
   (e) "chilly"
2.8 Which of the following words have intersection?
   (a) "arid"
   (b) "dry"
   (c) "wet"
   (d) "humid"
2.9 Which of the following words imply union?
   (a) "sand"
   (b) "boulder"
   (c) "alluvium"
   (d) "gravel"
   (e) "concrete"
   (f) "river"
2.10 Which of the following words imply a more or less similar range of numerical values?
   (a) "noon"
   (b) "almost a dozen"

      (c) "human tallness"

      (d) "months"

      (e) "teenager"

2.11 What are the differences among the following words?

      (a) "cylinder"

      (b) "disk"

      (c) "stick"

2.12 Briefly explain the differences between the following words:

      (a) "vague"

      (b) "ambiguous"

      (c) "incomplete"

      (d) "uncertain"

      (e) "fuzzy"

2.13 Arrange the following words in approximate numerical order:

      (a) "boulder"

      (b) "silt"

      (c) "gravel"

      (d) "clay"

      (e) "sand"

2.14 Arrange the following words in ascending order:

      (a) "cat"

      (b) "ant"

      (c) "giraffe"

      (d) "elephant"

      (e) "bee"

      (f) "mosquito"

2.15 Provide an index of numbers collectively for the following words:

      (a) "arid"

      (b) "dry"

      (c) "desertification"

      (d) "desert"

2.16 What are the words that classify "temperature" into at least four classes?

2.17 Give at least three imprecise hydrological phenomena that can be explained in words and sentences for non-experts.

2.18 Which scales are dominant in hydrological sciences, and are they certain, random, or fuzzy?

      (a) "micro"

      (b) "macro"

      (c) "large"

2.19 Which of the following words best explain cake cooking?

      (a) "brown color"

      (b) "60°C to 70°C"

      (c) "almost 65°C"

2.20 Order the following words numerically on a "rainfall" linguistic variable axis:

      (a) "dry"

      (b) "semi-dry"

      (c) "arid"

      (d) "humid"

      (e) "wet"

2.21 Explain the following glasses, which are half-full or half-empty, by CL and FL.

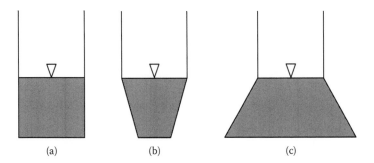

(a)   (b)   (c)

3 Fuzzy Sets, Membership Functions, and Operations

3.1 GENERAL

Fuzzy sets were introduced by Zadeh (1965) as a mathematical way to represent vagueness in linguistics that can be considered a generalization of classical (crisp) set theory. One of the biggest differences between crisp sets and fuzzy sets is that the former always have unique membership functions (MFs) in the form of a rectangle over the member elements, whereas every fuzzy set has infinite MFs that may represent it. In a broad sense, any field such as hydrology can be fuzzified and hence generalized by replacing a crisp set in the target field by the concept of fuzzy sets.

To fully understand fuzzy sets, one must understand traditional (crisp) sets. The rather abstract concept of a set forms a fundamental building block of modern mathematics and logic. Traditional mathematics and logic assign a membership degree (MD) of 1 to items that are members of a set and 0 to those that are not. This is the dichotomy principle. Such a strong principle inevitably runs into philosophical problems. Fuzzy set theory offers a resolution to such problems. More importantly, it offers a logic that closely imitates the human thought process by allowing for approximate reasoning and vagueness. It allows a proposition to be neither "fully true" nor "fully false" but "partly true" and "partly false" to a given degree. It is common to restrict these MDs to the real inclusive interval [0,1].

Fuzzy sets differ from classical or crisp sets in that they allow partial or gradual MDs, which is easily appreciable by looking at the difference between a conventional (crisp) set and a fuzzy set. A fuzzy set is almost any condition for which there are words such as *rare rainfall, intensive rainfall, high discharge, almost dry, highly wet, almost saturated layer, low groundwater velocity, slightly humid, rather porous*, etc., where the condition can be given any value (MD) between 0 and 1. For example, a discharge is 670 m^3/sec may be one of the highest amounts that a hydrologist has ever experienced in his or her life and it may be rated as 0.97. This line of reasoning can go on indefinitely, that is, rating a great number of things between 0 and 1.

It was already explained in the previous chapters that some words related to the problem at hand imply numerical values and hence it is now time to identify such numerical content of words through fuzzy sets. Each word is a collection of objects in a categorization and, likewise, each set is the collection of these objects with belongingness values or MDs as a measure of their belongingness to the set. The MD is a continuous allocation of objects between 0 and 1 inclusive. Crisp sets of CL have two mutually exclusive counterparts, each with a belongingness value of either

0 or 1. However, fuzzy sets are not mutually exhaustive but overlap each other and hence the MD of each object to a set is partial. It is well known that crisp sets have characteristic functions in the form of rectangles, whereas fuzzy sets do not have sharp endings but smooth transitions that give rise to neighboring sets that interfere with each other to a certain extent. Consequently, the characteristic functions of fuzzy sets are not in the form of rectangles but in forms of triangles, trapeziums, bell shapes, etc., which have at least one object in the set with MD equal to 1. Such characteristic functions are referred to as MFs.

In this chapter, and after the presentation of fuzzy set properties and MFs with their mathematical expressions, logical set operations are explained in detail.

3.2 CRISP AND FUZZY SETS IN HYDROLOGY

In mathematics, the concept of a set is simple but very important. A set is simply a collection of things (values, members, items, elements). The things can be anything one wants, such as numbers, names of drainage basins, etc. In the classical concept, things either belong to the set or do not belong, similar to the idea in Aristotlean logic where the statements are either true or false.

The collections of distinctive objects constitute the crisp (classical) sets, which dichotomize the elements of the universe of discourse into two adjacent and mutually exclusive groups. Whatever the objects, they are classified verbally as members or non-members of the set. Hence, the crisp sets are represented by the characterization of each element into two distinct and mutually exclusive and exhaustive groups with characteristic function (CF). Because there are two numerical characteristic values (CVs), as 1 for members and 0 for non-members, the CF appears as a rectangle that covers the whole domain of variation with the same height that is equal to 1 as shown in Figure 3.1. This figure represents the fuzzy word discharge Q, of some river based on the measured values, q, versus the CV, $\mu_Q(q)$. Because *discharge* measurements are confined between the minimum, q_{min}, and the maximum, q_{max}, the CF has a range between these two values.

Non-member regions surround discharge members on both sides, and the boundaries between the member and non-member regions are rigid and sharp. The universe of discourse is crisp. Hence, the CF takes on values as either 0 or 1, {0, 1}, and

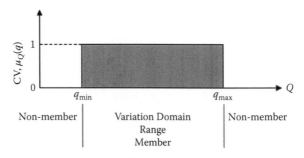

FIGURE 3.1 Characteristic function of discharge set.

is defined such that $\mu_Q(q) = 1$ if q is a member within the range. This statement can be shown symbolically as:

$$\mu_Q(q) = \begin{cases} 1 & \text{if and only if} \quad q \in Q \\ 0 & \text{if and only if} \quad q \notin Q \end{cases} \tag{3.1}$$

where $q \in Q$ ($q \notin Q$) means that q belongs (does not belong) to set Q. Figure 3.1 cannot be representative of natural hydrological events due to the following reasons:

1. According to Figure 3.1, all the observed discharge values have equally likely belongingness (CVs) in the discharge set.
2. "High" and "low" discharge values have the same emphasis—to be inclusive within the discharge set as the representative discharges. This is tantamount to saying that there is no distinction (granulation), for instance, between extreme discharges ("flood" and "drought") and the average discharge value, which lies about in the middle of the range.
3. Rareness of some discharge values cannot be distinguished from other discharges.

Fuzzy thinkers say that all things are matters of degree (not 0 and 1 only), and also reduce black-white logic (CL) and mathematics to special limiting cases of gray relationships (FL). Mathematically, fuzziness means multi-valence so that multi-valued fuzziness corresponds to degrees of indeterminacy or ambiguity, partial occurrence of events or relations.

Logically, in order to get rid of some of the crisp drawbacks, Figure 3.1 can be rendered into Figure 3.2 with gradually varying CVs. This implies that not all the discharges have the same belongingness value in the same set. In this case, the belongingness value (CV) is referred to as the MD and its function is the MF, which is a shape that defines how each point in the universe of discourse is mapped to a value between 0 and 1, inclusive. In Figure 3.2, the discharge MF is taken as a triangle because it is the simplest shape that translates all of what has been said about the unequal belongingnesses within each fuzzy word. It means that high and low

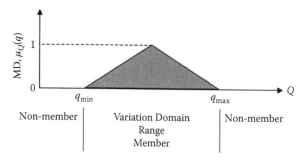

FIGURE 3.2 MF of discharge set.

discharges have comparatively less MD than the medium values. In this way, the crisp value appearance representative in Figure 3.1 as rectangle is fuzzified into a triangular MF. Hence, fuzzification means allocation of different MDs to values within a fuzzy word.

A fuzzy set is a group of anything that cannot be precisely defined. For instance, a fuzzy set of "discharges" gives answers to questions such as "How big is a discharge?" "Where is the dividing value between "low" and "medium" discharge?" "Is a 50 m^3/sec discharge a "low" or "moderate discharge" quantity?" "How about 75 m^3/sec?" "What about 156 m^3/sec?" The assessment should take into consideration the expert view of the hydrologist based on the catchment as well as meteorological conditions. Other examples of fuzzy sets are "hazardous flood," "short duration drought," "warm days," "high hydraulic head," "small drainage area," "medium viscosity," "brackish water," etc. For an analysis, it is necessary to have a way to assign some rational value to intuitive assessments for individual elements of a fuzzy set. Human fuzziness must be translated into numbers that can be used by a computer. This can be done by the assessment of a value from 0 to 1. For the question of "How severe is the rainfall?," the human might rate it at 0.3 if the rainfall is low. The hydrologist might rate the severe rainfall at 0.9, or even at 1.0 MD, if it is in winter season. These perceptions are fuzzy, just intuitive assessments, not precisely measured facts.

One must not confuse the MF with the probability distribution function (PDF), where the data range is divided first into non-overlapping, mutually exclusive, and adjacent sub-classes and then the relative frequency of each sub-class is calculated. The plot of these relative frequencies versus the discharge is referred to as a histogram, which may have the shape as depicted in Figure 3.3, where there are seven sub-classes that are mutually exclusive and each one represents the percentage (probability) of data that lies within each sub-class shown on the vertical axis as the relative frequency. By definition, each relative frequency is less than 1 but their summation is equal to 1. This is the main reason why each sub-class relative frequency is less than 1 in Figure 3.3, and has the same interpretation similar to Figure 3.1 with adjacent rectangle functions for each sub-class.

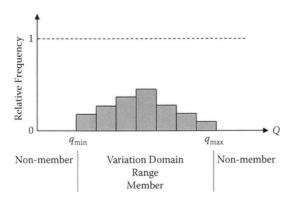

FIGURE 3.3 Histogram function of discharge set.

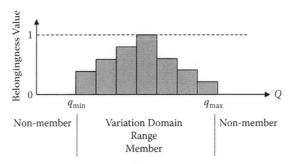

FIGURE 3.4 Histogram function of discharge set.

However, in the sub-class with the maximum relative frequency does not have the relative frequency value equal to 1; therefore, various classes cannot be regarded as proper sets. It is possible to transform Figure 3.3 such that the maximum relative frequency becomes equal to 1 as in Figure 3.4, which now has the property of the maximum belongingness (CV or MD) equal to 1.

Figure 3.4 can be regarded as a special case of Figure 3.3 where there is a change in the MD in the form of a step-wise reduction on both sides. None of the sub-classes have the property of crisp set except the one in the middle with its belongingness value being equal to 1 (similar to Figure 3.1).

The above arguments indicate that the sub-classes in a histogram cannot be appreciated as sets but only probability (relative frequency) representations, and the boundaries between sub-classes are still sharp and rigid. This implies that any element of discharge cannot be a member simultaneously in two or more sub-classes but should remain within a single sub-class only. The sub-classes have numerical labels as the lower and upper boundaries, as explained in any textbook on statistics (Benjamin and Cornell, 1973).

If the numerical classification is thought of as a linguistic variable, then the discharge domain can be categorized by words such as *low*, *medium*, and *high*. Hence, three subsets of the overall "discharge" set are considered with their belongingness values that must be either 0 or 1 (crisp subsets) as shown in Figure 3.5.

In this figure, "low," "medium," and "high" discharges are shown separately and collectively as crisp sets, and therefore each subset has its own domain of members and non-members.

On the other hand, a fuzzy set introduces vagueness by eliminating sharp boundaries that divide members from non-members. In practice, computers cannot distinguish "low" discharges from the "medium" or "medium" from the "high" values without crisp guidance. In verbal or linguistic (word) expressions, there is always overlapping between adjacent subsets, such as "low," "medium," and "high." Thus, the transition between full membership and non-membership is with boundary degradation in a gradual manner rather than abrupt boundaries. Hence, consideration of these two last sentences leads to the fuzzy subsets of discharge as

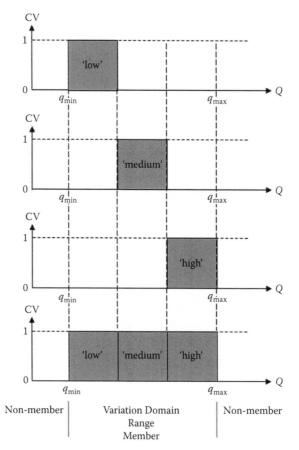

FIGURE 3.5 Crisp subsets of discharge set.

presented in Figure 3.6. Here again, triangular and trapezium MFs are used but they can have any shape, provided that there is at least one value with MD equal to 1 (see Section 3.4).

In this manner, the fuzzy word *discharge* is first categorized into three sub-fuzzy words—namely, *low*, *medium*, and *high*—and then each one of these adjectives are fuzzified. This is the basis of the fuzzification operation in fuzzy methodology applications (Chapter 6). Fuzzification means allocation of a suitable MF to each adjective word. A significant point is to notice that the rightmost and leftmost MFs are in trapezium forms. If they are left as triangles, their interpretation appears against the FL principles because any discharge less than q_{min} will have MD less than 1, which is implausible because there is no other MD on the left-hand side apart from "low" MF (Chapter 7).

After all, a universe of discourse is the concept that a linguistic variable describes. It could be many quantitative measurements, such as the length of a lineament, or the amount of discharge, etc. Consider a universe of discourse X with elements x. The crisp set A of X is defined by the characteristic function $f_A(x)$ of A, $f_A(x)$: $X \rightarrow [0,1]$, where $f_A(x) = 1$ if $x \in A$, or $f_A(x) = 0$ if $x \notin A$ (Equation (3.1)). However, a fuzzy set B

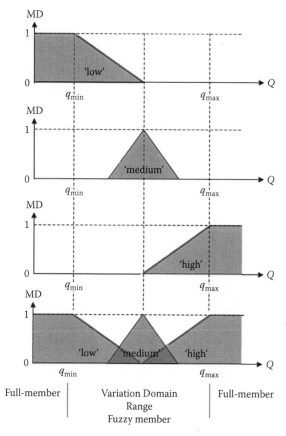

FIGURE 3.6 Fuzzy subsets of discharge set.

over the same universe of discourse X is defined by an analogous MF, $\mu_B(x)$, as $\mu_B(x)$: $X \rightarrow [0, 1]$, where $\mu_B(x) = 1$ if x is completely in B, $\mu_B(x) = 0$ if x is completely outside B, and $0 < \mu_B(x) < 1$ if x is partly in B. The $\mu_B(x)$ represents the MD of an element x in the fuzzy set B. If B includes n elements $(x_1, x_2, \ldots, x_n)$ with corresponding MDs as $\mu_B(x_1), \mu_B(x_2), \ldots, \mu_B(x_n)$, then the open notation for B can be expressed in terms of its elements and MDs as:

$$B = \left\{ \frac{\mu(x_1)}{x_1} + \frac{\mu(x_2)}{x_2} + \cdots + \frac{\mu(x_n)}{x_n} \right\} \tag{3.2}$$

Considered a crisp set, the elements of a fuzzy set constitute the support ("S") of this set as "S" = $\{x_1, x_2, \ldots, x_n\}$. In Equation (3.2), the plus signs do not imply the summation operation but rather convenience for fuzzy set element separation. The division sign in each element separates MD in the nominator from the element descriptor in the denominator. The x values in the set are in increasing order $(x_1 \leq x_2 \leq \cdots \leq x_n)$.

For example, the $B = \{0.25/1 + 1.0/2 + 0.1/3\}$ fuzzy set has three elements (1, 2, and 3) that belong to this set with MDs 0.25, 1.0, and 0.1, respectively. The corresponding crisp set can be written in general as $A = \{1.0/1 + 1.0/2 + 1.0/3\}$ or, because the MDs are all equal to 1, it is the support of the fuzzy set as $A = \{1, 2, 3\}$.

Example 3.1

Let us assume that in a drainage basin, past records indicated that the flood discharges can be categorized into 11 discrete groups with representative values as $Q = \{50, 100, 150, 200, 250, 300, 350, 400, 450, 500, 550\}$ m³/sec. Consider that these values are categorized into three fuzzy subsets as "high," "medium," and "low" floods with the MDs as allocated in Table 3.1. According to this table, "high" flood discharges can be found in a crisp manner as the support of the "high" subset of flood discharge as:

"high" = {250, 300, 350, 400, 450, 500, 550}

Likewise, the supports (crisp subset discharges) for "medium" and "low" flood discharges are:

"medium" = {150, 200, 250, 300, 350, 400, 450}

and

"low" = {50, 100, 150, 200, 250, 300, 350},

respectively. Herein, the ordered pairs for the "low" flood fuzzy subset can be arranged according to Equation (3.2) as:

"low" = {1.0/50 + 1.0/100 + 0.8/150 + 0.6/200 + 0.4/250 + 0.2/300 + 0.1/350}

TABLE 3.1
Flood Discharge MD Allocations

Flood Discharge (m³/sec)	"high"	"medium"	"low"
50	0	0	1.0
100	0	0	1.0
150	0	0.2	0.8
200	0	0.6	0.6
250	0.1	0.9	0.4
300	0.3	1.0	0.2
350	0.5	0.9	0.1
400	0.7	0.6	0
450	0.8	0.2	0
500	0.9	0	0
550	1.0	0	0

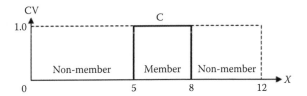

FIGURE 3.7 Crisp set.

A traditional (crisp) set can formally be defined as follows:

1. A crisp subset C of a set X is a mapping from the elements of X to the elements of the set {0,1}.
2. The mapping is represented by one ordered pair for each element X, where the first element is from the set X, and the second element is from the set {0,1}. The MD value "0" represents non-membership (non-belongingness), while the value "1" represents full-membership (belongingness).

Essentially, this indicates that an element of the set X is either a member or a non-member of the subset C. There are no partial members in traditional sets. Consider a set X that contains all the real numbers between 0 and 12, and a subset C of the set X that contains all the real numbers between 5 and 8. Subset C is represented in Figure 3.7, where the interval on the X-axis between 5 and 8 has CV equal to 1. One should notice that the crisp set appears in the form of a rectangle with height equal to 1, and the lower and upper boundaries are crisp (sharp).

On the other hand, a fuzzy set has elements with different MDs, which can formally be defined as follows:

1. A fuzzy subset F of a set X can be defined as a set of ordered pairs. The first element of the ordered pair is from the set X, and the second element from the ordered pair is from the interval [0,1].
2. The value "0" ("1") is used for complete non-belongingness (complete belongingness), whereas other values in between are used to represent partial belongingness.

These two points imply that the form of a fuzzy set cannot have sharp (crisp) boundaries and hence the fuzzy set that corresponds to Figure 3.7 can take the form as in Figure 3.8.

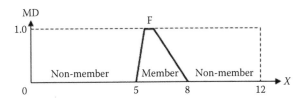

FIGURE 3.8 Fuzzy set.

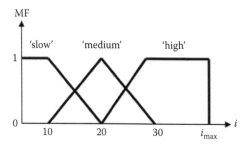

FIGURE 3.9 Intensity linguistic variable and fuzzy subsets.

For instance, the intensity, i, of a storm rainfall can take values in the interval $[0, i_{max}]$, where i_{max} shows the maximum possible intensity. It is possible to subdivide this linguistic variable into a few fuzzy subsets such as "slow," "medium," and "high," which are shown in Figure 3.9.

In a more formal way, a linguistic variable is characterized symbolically as a quadruple (V, L, D, R) where

1. V symbolizes the name of the variable, such as "intensity," "evaporation," "rainfall," runoff," "seepage," etc.
2. L represents the set of linguistic values that that linguistic variable can take. These may be "slow," "medium," and "high."
3. D is the actual physical domain in which the linguistic variable assumes its values.
4. R is the semantic rule that relates each linguistic value in L with a fuzzy set in D. This corresponds to MFs and particularly MD for a given member.

Unlike in CL, where there are only two logical choices—"wet" and "dry," "high" and "low"—fuzzy theory can accept indefinite choices (multi-value logic). In CL, for example, there is only "wet" membership and "dry" membership sets. Any one year can only be a member in a "dry" year or not at all. Hence, this year cannot be a member of two or more sets at the same time. In reality, during persistent drought spells, some years are definitely members of the "dry" period only. During the transition periods (crossing-points), however, between dry and wet spells, some years may be a member of a "rather dry" spell, and at the same time they may also have a tendency toward a "wet" spell with lower degrees of acceptance less than 1. However, classical logic uses a clear-cut boundary, for example, "dry," which may be defined for those years with less than a truncation level, X_0 (Figure 3.10).

There has been a long debate on the membership of "dry" close to the truncation level. Those close to the truncation level are neither a "wet" nor "dry" spell according to fuzzy definition. Instead of a clear-cut boundary, there are MDs in the fuzzy sets of "wet" and "dry." In fuzzy concept, any spell may be defined in three categories:

1. Those whose values are definitely well over the truncation level have constant MD equal to 1.0 in the "wet" fuzzy set.
2. Those whose values are well below the truncation level have constant MD equal to 1.0 in the "dry" fuzzy set.

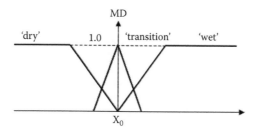

FIGURE 3.10 Sets of "wet," "transition," and "dry."

3. Those whose values are around the truncation level have MD decreasing from 1.0 to 0 toward the "wet" and "dry" fuzzy sets, which is referred to as the "transition" fuzzy set.

The MD variations of the "dry" spell can be represented logically as in Figure 3.10, which implies that as the value moves away from the truncation level toward smaller values, it becomes more and more dry, and finally, absolute dryness is the virtual state of the hydrological phenomenon with MD equal to 1.0.

Example 3.2

If the drought threshold value in an area is taken as $Q = 17.5$ m³/sec, then the discharge values greater than this value will have the following crisp set:

$$\mu_Q(q) = \begin{cases} 0 & \text{drought (dry)} & \text{if and only if} \quad Q \le 17.5 \text{ m}^3/\text{sec} \\ 1 & \text{not drought (wet)} & \text{if and only if} \quad Q \ge 17.5 \text{ m}^3/\text{sec} \end{cases} \tag{3.3}$$

which is shown in Figure 3.11a. In this manner, the variation domain of the "discharge" variable is categorized crisply into two adjacent but mutually exclusive subsets.

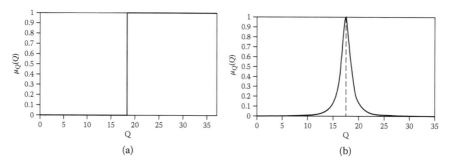

FIGURE 3.11 (a) CF and (b) MF.

In the discussions above, the fuzzification of the fuzzy word *discharge* is achieved by the triangular and trapezium MFs because these are the simplest logical shapes for the requirement of MD granulation. It is also possible to attach MFs through mathematical reasoning of the fuzzy statements. For instance, the fuzzy statement threshold "close to 17.5 m³/sec" gives the impression that one should search a functional shape that should have its MD equal to 1 at 17.5 m³/sec and decrease in the MD continuously as the discharge value moves away from 17.5 m³/sec on both sides. The previous sentence can be translated into a mathematical MF, $\mu_Q(q)$, rationally using the following guideline information:

1. As $q \rightarrow 17.5$ m³/sec, $\mu_Q(q) \rightarrow 1$. This can be achieved after some thought as:

$$\mu_Q(q) = \frac{1}{1 + (q - 17.5)}$$

2. By definition, the MF cannot assume negative degrees, $[0 \leq \mu_Q(q) \leq 1$, which is not satisfied by this expression, because, say, for $q = 10$ m³/sec, it yields $\mu_Q(q) = -1/6.5 < 0$ and cannot be acceptable. To get positive MDs whatever the "discharge" value, the difference term in the denominator must be positive; and the most convenient way is to take the square of this term, which yields:

$$\mu_Q(Q) = \frac{1}{1 + (Q - 17.5)^2} \tag{3.4}$$

with its graphical presentation in Figure 3.11b.

It is also possible to generalize the expression threshold "close to 17.5 m³/sec" by considering a factor α in Equation (3.4) as:

$$\mu_Q(Q) = \frac{1}{1 + \alpha(Q - 17.5)^2} \tag{3.5}$$

which presents different MFs for the same phenomenon. It is logical that the greater the certainty, the narrower the function, as in Figure 3.12. An increase in the α value gives rise to an increase in the precision and a reduction in the vagueness (i.e., fuzziness) (see Section 3.6).

This example shows that fuzziness is a type of impression that stems from a grouping of elements into classes that do not have sharply defined boundaries. Although the MD of each element in the fuzzy set is less than 1, their summation must not be expected to equal 1 as in probability theory.

Example 3.3

Here is an example describing a fuzzy set of "potable" waters by considering a total dissolved solids (TDS) value measured in parts per million (ppm). It is stated in the literature that a TDS value less that 500 ppm is the limit for "potable" water. If one uses this strict interval to define "potable" water, then water with 500 ppm is

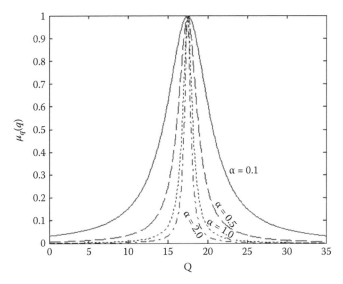

FIGURE 3.12 Different precisions of fuzzy expression "threshold close to 17.5 m³/sec."

still "potable" (still a member of the set). For any TDS value greater than 500 ppm, for example, 500.1 ppm, according to CL it is not "potable" (i.e., not a member of the potable set). How can one remedy this problem?

It is necessary to relax the boundary between the crisp separation of "potable" and "non-potable" classes. This separation can easily be relaxed by considering the boundary between "potable" and "non-potable" as fuzzy (Figure 3.13). Notice in the figure that waters whose TDS values are greater than 0 ppm and less than 500 ppm are complete members of the "potable" water set (i.e., they have MDs equal to 1). Also note that waters with TDS values greater than 500 ppm and less than 700 ppm are partial members of the "potable" water set. For example, water with 600 ppm TDS would be "potable" to the MD of 0.5. Finally, waters whose TDS values are greater than 700 ppm are non-members of the "potable" set.

Example 3.4

Consider that there are three soil types: "fine" ("F"), "medium" ("M"), and "course" ("C"). These types, separately, represent crisp sets. For example, true fine is a

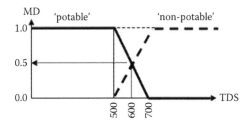

FIGURE 3.13 "Potable" and "non-potable" water MFs.

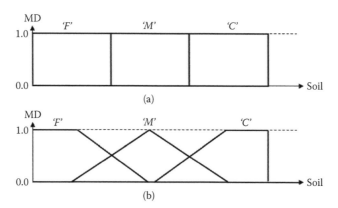

FIGURE 3.14 Soil classification sets: (a) crisp and (b) fuzzy.

non-member of "true medium" and of "true course"; "true medium" is a non-member of "true fine" and of "true medium"; "coarse" is a non-member of "true fine" and of "true medium." There is a crisp boundary between these primary soil types (Figure 3.14a).

It is possible to mix these types with varying amounts of the true grain sizes, resulting in different types of soil mixtures. For example, mixing "true fine" with "true medium" in equal portions of each will result in a mixture with an MD of 0.5 in "true fine" and 0.5 in "true medium." Different amounts of "true fine" and "true medium" will result in various MDs for the mixture. The different mixtures represent the fuzzy boundaries between "true fine" and "true medium" as in Figure 3.14.

Example 3.5

Let U denote the universal set, which includes all the possible elements of a variable. For instance, the universal set of rainfall magnitudes in the Arabian Peninsula deserts include at first instant positive quantities between 0 mm and a known maximum value, say, 50 mm, from historical records. This restricts the variability domain of the region's rainfall amounts to an observed maximum value, and therefore it is a crisp universal set. However, if the maximum is fuzzified by saying "around" or "approximately," then the universal set is implied as a fuzzy set. Although one can define the crisp set by listing its elements, this will be very tiring and practically impossible. Instead, it is better to describe any set by its rule, which implies meeting some conditions. For instance, a rainfall set R can be represented in the rule method as:

$$R = \{r \in U | r \leq 50\} \tag{3.6}$$

This means that the discourses of set R on the universal set U are rainfall amounts r that are less than or equal to 50 mm. Another way of representing the elements of a set is through the CF, which introduces a zero-one MF. Other names

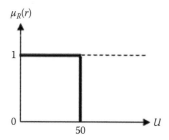

FIGURE 3.15 Characteristic function.

for the CF are discrimination (member and non-member) or indicator functions. If the MF of a universal set is shown by $\mu_R(r)$ similar to Equation (3.1), then

$$\mu_R(r) = \begin{cases} 1 & \text{if} \quad r \in R \\ 0 & \text{if} \quad r \notin R \end{cases} \tag{3.7}$$

Consideration of the previous expression with the definition in Equation (3.6) leads to the graphical representation of a crisp set A in the universal set U as in Figure 3.15.

Example 3.6

Now consider the sub-basins in some drainage basin, which is the universe of discourse, U. One can distinguish different sets in U according to the sub-basins. Such a classification is given in Figure 3.16, where the whole drainage basin is considered first as "humid" sub-basins and "non-humid" sub-basins. Another division can be made according to drainage orders as 2nd, 3rd, 4th, and others.

It is possible to think about a set as all sub-basins in U that are 3rd order. This can be written as

$$F = \{d \in U \mid d \text{ has 3rd order streams}\} \tag{3.8}$$

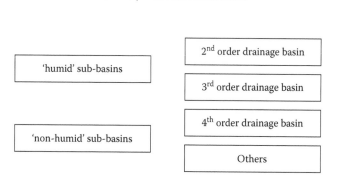

FIGURE 3.16 Euphrates basin subsets.

where d represents drainage basin. It can be written in terms of CF as

$$\mu_d(x) = \begin{cases} 1 & \text{if} \quad q \in U \text{ and } d \text{ has 3rd order stream} \\ 0 & \text{if} \quad q \notin U \text{ and } d \text{ hasnot 3rd order stream} \end{cases}$$

There arises a difficulty when defining a set in U according to whether the drainage basin is "humid" or "not." This is due to the fact that the drainage basin might not be entirely "humid" or "non-humid" but may have a common feature in between. Many hydrologists will agree that some of the drainage basins cannot be described crisply. So, how to deal with this kind of problem is the main question. This indicates that in the given example, some sets do not have clear boundaries, as the classical sets require. This difficulty can be solved by fuzzy sets, which have transitional (non-crisp) boundaries, that is, boundaries that overlap between two or more subsets.

Example 3.7

It is now possible with fuzzy set conception to identify the subsets of sub-drainage basins of some areas. Let this set, denoted by PR, be a fuzzy set according to the percentage of sub-basins in the river basin. It can be expressed with the MD as:

$$\mu_{PR}(x) = p(x) \tag{3.9}$$

where $p(x)$ is the percentage of the sub-drainage basins within the river basin. This percentage can have, in general, any value between 0 and 100 percent. For instance, some far away rivers do not have full MDs of belongingness to this river basin. Furthermore, sub-drainages outside the river basin, NT, are the complementary event, and therefore its MD is given as,

$$\mu_{NR}(x) = 1 - p(x) \tag{3.10}$$

This is the crisp definition of the percentages. In the case of vagueness, the MFs may be considered as in Figure 3.17.

In this manner, the first fuzzy set operation as simple complementary is achieved as shown in Figure 3.17 (see Section 3.7.5).

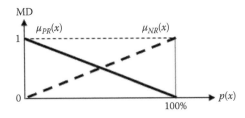

FIGURE 3.17 Fuzzy subsets and MFs.

Example 3.8

If the relative humidity is taken as a fuzzy discourse universe variable, its values are confined between 0 and 100 percent. The "low" and "high" relative humidity subsets can be expressed with integral notation of fuzzy sets as:

$$\text{"low"} = \int_0^{40} 1/h + \int_{40}^{100} \left[1 + \frac{1}{\left(\dfrac{h-40}{4}\right)^2} \right]^{-1} \Big/ h \tag{3.11}$$

and

$$\text{"high"} = \int_{40}^{100} \left[1 + \frac{1}{\left(\dfrac{h-40}{4}\right)^2} \right]^{-1} \Big/ h \tag{3.12}$$

Example 3.9

In the case of discrete elements, the fuzzy sets can be expressed using the notation as in Equation (3.2). Let X include integers from 1 to 12 as in Figures 3.7 and 3.8; then the fuzzy subset "several" can be written as:

$$\text{"several"} = \left\{ \frac{0.4}{3} + \frac{0.8}{4} + \frac{1.0}{5} + \frac{1.0}{6} + \frac{0.8}{7} + \frac{0.5}{8} \right\} \tag{3.13}$$

which shows 5 and 6 as complete members of given integer numbers belong to the fuzzy set "several" with degree 1; 4 and 7 with degree 0.8; 8 with degree 0.5; and finally, 3 with degree 0.4.

Example 3.10

Consider a fuzzy set groundwater velocity with fuzzy members "slow," "medium," and "fast." The expert view of a groundwater hydrologist might attach the MDs to each fuzzy wording as in Table 3.2. This example indicates at this stage that the fuzzy variable "velocity" as a linguistic variable and universal set has three linguistic adjectives—"slow," "medium," and "fast"—as subsets. Thus far, even the non-hydrologist can logically suggest these three categories but the significance of a groundwater hydrologist becomes important for the attachment of a plausible

TABLE 3.2
Groundwater Velocity Fuzzification

Velocity (m/day)	MDs		
	"Slow"	"Medium"	"Fast"
1.0	1.0	0.0	0.0
2.0	0.8	0.0	0.0
3.0	0.6	0.4	0.0
4.0	0.4	0.6	0.0
5.0	0.2	0.8	0.0
6.0	0.0	1.0	0.0
7.0	0.0	1.0	0.0
8.0	0.0	1.0	0.0
9.0	0.0	1.0	0.0
10.0	0.0	1.0	0.0
11.0	0.0	0.8	0.2
12.0	0.0	0.6	0.4
13.0	0.0	0.4	0.6
14.0	0.0	0.0	0.8
15.0	0.0	0.0	1.0

variation domain for the groundwater velocity, its subdivision into three overlapping categories, and last but not least, the attachment of an MD (fuzzification) for each subset.

It is necessary to find some way to go back and forth between the description of velocity in numbers and the description in words. This is done by defining MFs as in Figure 3.18.

Different hydrologists would have different perceptions about how velocity should be classified as "slow," "medium," and "fast," depending on their working environment. It must be noticed that leaving a good deal of overlap on the adjacent MFs is a very good idea because it lends robustness to system reasoning. MFs need not be made up of straight lines as in Figure 3.18; if necessary, they also can have curvilinear shapes (see Section 3.4).

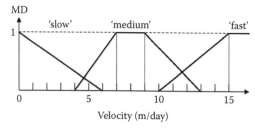

FIGURE 3.18 Groundwater velocity fuzzy sets.

Other useful fuzzy sets are not descriptions of numbers, but of categories (in a broad sense). These fuzzy sets are not word descriptors of "how much," but are answers to "what is it?" For example, in the case of a rainfall intensity fuzzy set, "dense" might have members "extensive," "medium," and "rare." Likewise, fuzzy set "channel" might have members such as "river," "stream," and "creek."

In fuzzy sets like "velocity" and "vegetation cover," it is very likely that two or more descriptors might have non-zero MDs at the same time. This is due to the fact that the hydrologists are seldom certain that one descriptor is entirely right and the others are wrong. These are ambiguities and, as mentioned above, they lend strength to the reasoning process. In other fuzzy sets like "channel," it is not possible for a channel to be both a creek and a stream. In this case, if more than one member has a non-zero confidence, there will appear a contradiction that needs to be resolved, often by examining more data or information.

3.3   FORMAL FUZZY SETS

As discussed, the fuzzy set can be thought of as the transformer of different discourses within each set into a map, which assumes values within the positive unit interval. For instance, if the rainfall intensity, I, is the name of the whole set and "small" rainfall intensity is one of the subsets, then as the smallness of intensity deviates from this peak value of 1, the fuzzy set starts to decrease in a monotonous manner, down to 0 on both sides (Figure 3.19).

In general, a fuzzy set has an infinite number of MDs that provide the maximum ability to adjust its utility according to any situation. Hence, it is possible to say that any crisp set or concept can be fuzzified in many different ways. For example, any basic concept in the hydrology domain can be fuzzified, including the algorithms, models, and procedures. This renders the whole calculation into a soft computing domain without any sudden transitions but with gradual variations. The fuzzification provides greater generality, higher expressive power, and an efficient ability to model real-world hydrologic problems. In the meantime, a methodology for exploiting imprecision tolerance can be developed, depending on

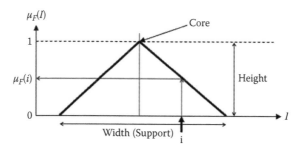

FIGURE 3.19 Fuzzy set with triangular MF.

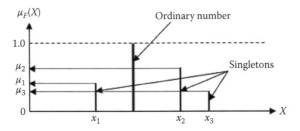

FIGURE 3.20 Singletons and ordinary number.

the availability of information, which is imprecise in many cases. These properties support tractable FL approaches with better robustness, less computation, and reduced solution costs.

The subset of the universe U in which the value of its MF is greater than zero $(\mu_F\{x_i\} \geq 0.0)$ is called the support of fuzzy subset F (Figure 3.19). A fuzzy subset whose support is only a single element is called a fuzzy singleton (Figure 3.20), which is represented as:

$$F = \left\{ \frac{\mu}{x} \right\} \tag{3.14}$$

where μ is the MD of element x (the only element). To distinguish it from an ordinary number (see Chapter 4), the MD of an ordinary number is always unity, while a fuzzy singleton may have MD less than unity.

Depending on the maximum MD equivalence to 1 or not, the fuzzy set is called normal [$\max\mu_F(x)$] = 1, or sub-normal [$\max\mu_F(x)$] ≠ 1, respectively (see Figure 3.21). As shown in Figure 3.19, the height of a normal fuzzy set is equal to 1. All members

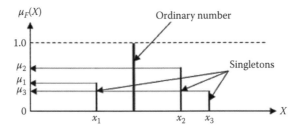

FIGURE 3.21 Fuzzy sets: (a) sub-normal and (b) normal.

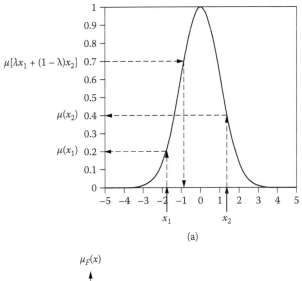

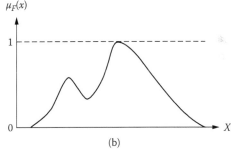

FIGURE 3.22 Fuzzy sets (a) convex and normal, (b) non-convex.

with MD = 1 in a fuzzy set constitute the core. Another property of fuzzy sets is their convexity, which is defined as:

$$\mu_F(\lambda x_1 + (1 - \lambda)x_2) \geq \min(\mu_F(x_1), \mu_F(x_2)) \tag{3.15}$$

where $x_1, x_2 \; \varepsilon \; U$ and $\lambda \; \varepsilon [0, 1]$. The fuzzy set in Figure 3.22a is a convex set, whereas a non-convex fuzzy set is shown in Figure 3.22b.

3.4 MEMBERSHIP FUNCTIONS

As already discussed in different parts of the previous chapters and sections, MFs are fuzzy sets that can be represented through mathematical expressions by taking into consideration the normality and convexity properties. In general, fuzzy sets can be of triangular, trapezoidal, Gaussian, or other form. In the transition regions, its MF parts can be linear, quadratic, or exponential, depending on the object of interest.

3.4.1 TRIANGULAR

The simplest MF has a triangular shape, which is a function of a support vector x and depends on three scalar parameters (a, b, and c) *as* shown in Figure 3.23. The MF is given by the following mathematical expression:

$$\mu(x) = \begin{cases} 0, & x \le a \\ \dfrac{x-a}{b-a}, & a \le x \le b \\ \dfrac{c-x}{c-b}, & b \le x \le c \\ 0, & c \le x \end{cases} \tag{3.16}$$

or, more compactly by

$$\mu(x) = \max\left[\min\left(\frac{x-a}{b-a}, \frac{c-x}{c-b} \right), 0 \right] \tag{3.17}$$

The parameters a and c locate the left and right limits of the support of the set and the parameter b locates the core.

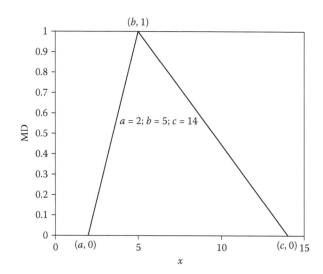

FIGURE 3.23 Triangular MF.

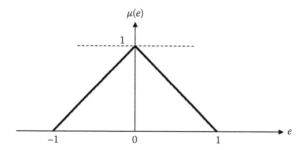

FIGURE 3.24 Triangular error membership function.

Example 3.11

In hydrology, many times the error percentage, e, is desired to be ±5 percent, which is a crisp requirement and does not make any distinction between the errors as long as they are between the lower (−5 percent) and upper (+5 percent) limits. However, "the error close to zero" is a requirement that is a fuzzy expression. In practical applications, it is more natural to demand such a linguistic condition. Although this cannot be expressed by classical sets or any mathematical procedures, the fuzzy concept can represent this linguistic requirement objectively.

It is possible to express the fuzzy set "the error close to zero" in a different and simpler manner by a triangular MF that varies between −1 and +1, where $a = -1$, $b = 0$ and $c = +1$ as in Figure 3.24.

Herein, the error of, say 2 and 0 has, respectively, MDs of 0 and 1. On the other hand, error $e = ±0.5$ has an MD of 0.5, etc.

3.4.2 TRAPEZIUM

The trapezoidal MF is a function of a support vector x and depends on four scalar parameters (a, b, c, and d), as shown in Figure 3.25. The mathematical expression of this MF is given by:

$$\mu(x) = \begin{cases} 0, & x \leq a \\ \dfrac{x-a}{b-a}, & a \leq x \leq b \\ 1, & b \leq x \leq c \\ \dfrac{d-x}{d-c}, & c \leq x \leq d \\ 0, & d \leq x \end{cases} \tag{3.18}$$

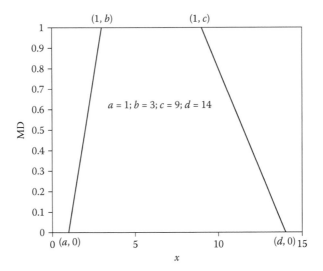

FIGURE 3.25 Trapezoidal MF.

or, more compactly, by:

$$\mu(x) = \max\left[\min\left(\frac{x-a}{b-a}, 1, \frac{d-x}{d-c}\right), 0\right] \tag{3.19}$$

The parameters a (b) and d (c) locate the left and right limits of the support. It is obvious that the triangular MF is a special case of the trapezoidal MF when $b = c$; and if $a = b = c = d$, then a fuzzy singleton is obtained. Furthermore, a, b, c, and d represent lower modal, left spread, upper modal, and right spread, respectively.

3.4.3 Sigmoid

The sigmoid MF depends on two parameters (a and c) and is given by:

$$\mu(x) = \frac{1}{1 + ae^{-(x-c)}} \tag{2.20}$$

The shape of the sigmoid MFs is shown in Figure 3.26 for different a and c parameters. Depending on the sign of the parameter a, the sigmoid MF is inherently open to the right or to the left, and thus is appropriate for representing concepts such as "very large" or "very negative." More conventional-looking MFs can be built by taking either the product or difference of two different sigmoid MFs.

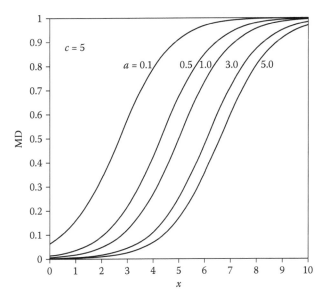

FIGURE 3.26 Sigmoid MF.

3.4.4 PROBABILITY DISTRIBUTIONS

All theoretical probability distribution functions (PDFs) are convex and hence can be adopted as an MF provided that the peak values are adjusted to have MD equal to 1. Figure 3.27 shows such adjusted curves, which are no longer PDFs but MFs. Among these, the most widely used one has the symmetric Gaussian shape in Figure 3.27a with a two-parameter mathematical form as:

$$\mu(x) = e^{-\frac{(x-m)^2}{2s^2}} \tag{3.21}$$

where s is the measure of spread around the value at m, which indicates the core (kernel)—that is, at m, the MD is equal to 1. If there are data, it might be taken as equal to a certain factor of the standard deviation of the variable concerned. Otherwise, it is just appreciation of the researcher to attach a convenient value. In Figure 3.27b, c, and d, other PDF type MFs are shown as Gamma, Weibull, and Gumbel curves, respectively.

Example 3.12

Another possible mathematical expression for the MF in Example 3.11 is the Gaussian MF with $m = 0$ and $s = 1$ (Equation (3.21)) as:

$$\mu_e(x) = e^{-\frac{1}{2}x^2} \tag{3.22}$$

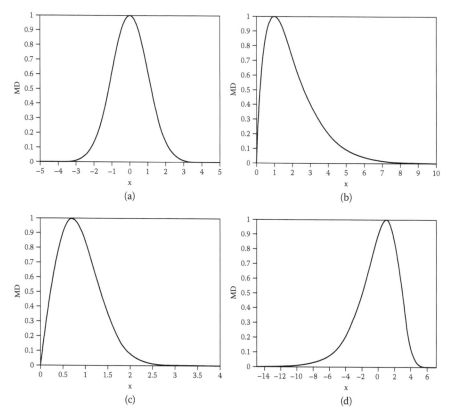

FIGURE 3.27 PDF MFs: (a) Gaussian, (b) Gamma, (c) Weibull, and (d) Gumbel.

The corresponding Gaussian MF is shown in Figure 3.28. Accordingly, the numbers −2, −1, 0, +1, and +2 belong to the fuzzy set of "the error close to zero" with MDs of 0.0183, 0.3679, 1.0000, 0.3679, and 0.0183, respectively.

3.4.5   TWO-PIECE GAUSSIAN

A two-piece Gaussian MF is a combination of two halves—namely, the leftmost and rightmost curves from the normal Gaussian curve as an MF. If these curves are shifted from each other to the left and right, respectively, and their core points are connected by a horizontal straight line, then a two-piece Gaussian MF is obtained, as in Figure 3.29. If the leftmost (rightmost) limb is specified by $s_1(s_2)$ and $m_1(m_2)$, the necessary and sufficient condition to obtain a properly defined two-piece Gaussian MF is that $m_1 > m_2$. Otherwise, the maximum MD value appears as less than 1.

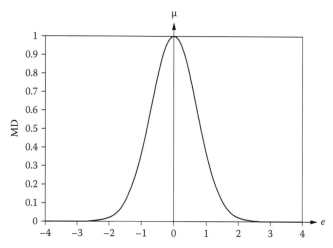

FIGURE 3.28 Gaussian error membership function.

3.4.6 GENERALIZED BELL SHAPE

The generalized bell MF depends on three parameters (a, b, and c) as given by:

$$f(x; a, b, c) = \frac{1}{1 + \left| \dfrac{x - c}{a} \right|^{2b}} \tag{3.23}$$

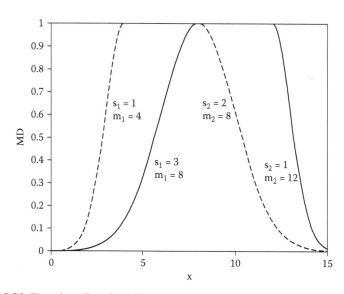

FIGURE 3.29 Two-piece Gaussian MF.

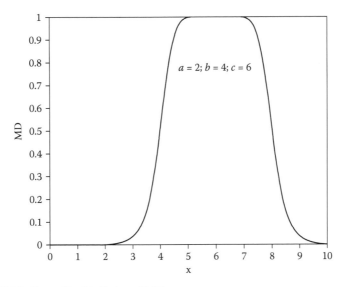

FIGURE 3.30 Generalized bell-shaped MF.

where the parameter b is usually positive. The parameter c locates the center of the curve (Figure 3.30).

3.4.7 S-Shape

The S-shape MF is defined by three parameters (a, b, and c) and its general shape is shown in Figure 3.31. The mathematical form of an S-shape MF is given as:

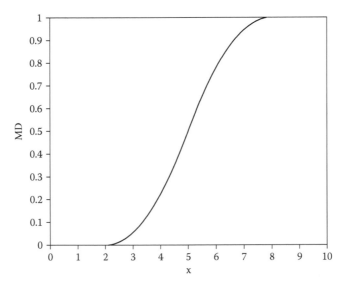

FIGURE 3.31 S-shaped MF.

$$f(x; a, b, c) = \begin{cases} 0 & \text{for} \quad x < a \\ 2\left(\dfrac{x-a}{c-b}\right)^2 & \text{for} \quad a < x < b \\ 1 - 2\left(\dfrac{x-a}{c-b}\right)^2 & \text{for} \quad b < x < c \\ 1 & \text{for} \quad x < c \end{cases} \tag{3.24}$$

The shape is very similar to the theoretical cumulative probability distribution (CPD) function curves for different PDFs, such as normal, Gamma, Weibull, Gumbel, and other curves, which can also be adopted as MF, if necessary.

3.4.8 Z-Shape

The Z-shape MF is the asymmetrical polynomial curve open to the left as shown in Figure 3.32 and its mathematical function has the following form:

$$f(x; a, b, c) = \begin{cases} 1 & \text{for} \quad x < a \\ 1 - 2\left(\dfrac{x-a}{c-b}\right)^2 & \text{for} \quad a < x < b \\ 2\left(\dfrac{x-a}{c-b}\right)^2 & \text{for} \quad b < x < c \\ 0 & \text{for} \quad x < c \end{cases} \tag{3.25}$$

This type of MF can be considered complementary to the CPD curve type of MF, as discussed in the previous section.

3.5 MEMBERSHIP FUNCTION ALLOCATION

The most common MF shape of is the triangle, although trapezoids and bell curves are also used, but the shape is generally less important than the number of curves and their placement in the variation domain support of a linguistic variable. In practical studies, three to seven MFs are generally appropriate to cover the support, or

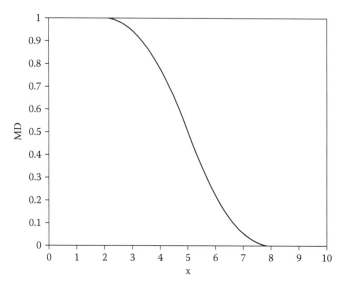

FIGURE 3.32   Z-shaped MF.

the "universe of discourse" in fuzzy jargon. The domain of variation can be broken down broadly by one of the following procedures:

1. *Subjective evaluation and elicitation:* As fuzzy sets are usually intended to model people's cognitive states, they can be determined from either simple or sophisticated elicitation procedures. At early stages, they are subject to simply drawings or otherwise different MFs are specified appropriate to a given problem. Experts in the problem area can suggest representative MFs. They may also be given a more constrained set of possible curves from which to choose. In more complex methods, users can be tested using psychological methods, even through initiative feelings.

2. *Ad hoc forms:* While there is a vast (hugely infinite) array of possible MFs, most actual fuzzy operations draw from a very small set of different curves; for example, simple forms of fuzzy numbers have triangular and trapezium shapes. This simplifies the problem, for example, to choosing just the central value and the slope on either side.

3. *Converted frequencies or probabilities:* As discussed in Sections 3.4.4, 3.4.7, and 3.4.8, information taken in the form of frequency histograms (PDFs) or CPD curves is sometimes used as the basis for constructing an MF (see Figure 3.27). There are a variety of possible conversion methods, each with its own mathematical and methodological strengths and weaknesses. However, it should always be remembered that MFs are not (necessarily) probabilities.

4. *Physical measurement:* Many applications of FL use physical measurements but almost none measure the MD directly. Instead, an MF is provided by another method, and then the individual MDs of data are calculated from it by fuzzification.

5. *Learning and adaptation:* The MD is the placement in the transition from 0 to 1 of conditions within a fuzzy set. If a particular dam on the scale has a rating of 0.7 in its capacity among whole dams in a region, then its MD in the fuzzy capacity set of dams is 0.7.

S application of FL procedures requires the preliminary determination of MF shapes of the input and output variables and writing down a suitable set of rules (Chapter 6). In many applications, the choice of MFs is achieved by a trial-and-error approach. As stated by Mendel (1995), prior to 1992, all FL systems reported in the open literature fixed the parameters of MFs somewhat arbitrarily; for example, the locations and spreads of MFs are chosen by the designer, independent of the numerical training data. Later, however, different methodologies such as artificial neural networks, genetic algorithms, simulated annealing method, and tabu search were employed in the modification and training of MFs (Ross, 1995; Şen, 2004).

The choice of MFs is usually problem oriented and are often determined heuristically and subjectively. During this selection, the hydrologist must consider readily available MFs such as triangular, trapezium, Gaussian, and S-shape. After choosing an MF shape, hydrologists must construct a set of MFs for all fuzzy linguistic terms for each input and each output variable. In a fuzzy model application, a set of MFs is normally adopted by considering the following points:

1. In general, the same shape of symmetric MF is selected for all fuzzy subsets (sub-words) of input and output linguistic variables.
2. These MFs are evenly distributed in the variation range of each variable, and adjacent MFs usually have a cross-point level equal to 0.5.
3. Different inputs and outputs may have the same or different shapes of MFs.

The evenly distributed triangular and trapezoidal MFs with four linguistic labels—"low" ("L"), "medium" ("M"), "high" ("H"), and "very high" ("VH")—are shown in Figure 3.33 for the linguistic variable *X*, variability range from 0 to 1.

Once the sub-groups within the data space are identifiable, it is straightforward to attach MFs to the problem at hand. For this purpose, there are two approaches, as subjective and objective groupings

3.5.1 Subjective Grouping

Heuristically chosen MFs cannot reflect the actual data distribution in the input and output space. However, if there is no data, they are the only sources of expert knowledge available. In the case of data availability, the MFs can be derived from the scatterplots (diagrams). For this purpose, the data are plotted in order to visualize the scatter and make visual interpretations. For instance, consider the scatter of points as in Figure 3.34. It is obvious at first glance that there are different categorizations such as A, B, and C.

Accordingly, the scatter diagram in Figure 3.34a (Figure 3.34b) shows two (three) subjective groupings without formal definition of the centers; however, the data are subdivided into two (three) parts, each with its specific location and dispersion

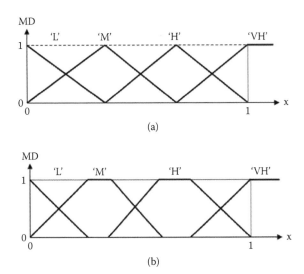

FIGURE 3.33 MFs: (a) triangular and (b) trapezium.

within the data variability space. In the former case, two groups have projections on the data axes without overlapping, but the latter has overlapping on both axes. The crisp variation domains of each group are shown by I, II, and III on the horizontal axis and by α, β, and γ on the vertical axis. However, at the time of MF attachment, these boundaries must be enlarged to intervene with each other partially. Projections of these groups onto the horizontal and vertical axes show the following features:

1. The number of fuzzy sets (MFs)
2. The variation domain (support) of each linguistic variable (X and Y)
3. The boundaries of each group where MDs are rather small
4. Approximate centers of each group where MDs can be taken as equal to 1

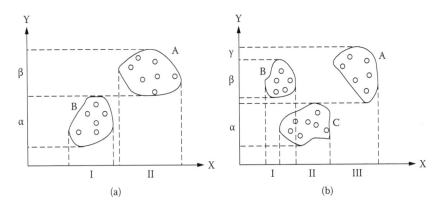

FIGURE 3.34 Heuristic categorization and MFs allocation.

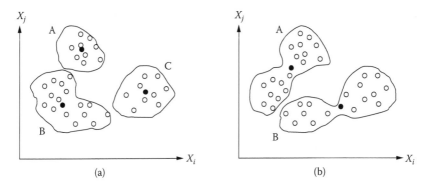

FIGURE 3.35 Clustering: (a) Euclidean distance and (b) angle-difference.

3.5.2 OBJECTIVE GROUPING

Apart from the visual scatter diagrams, there are also objective clustering methodologies that depend on the resemblance or dissemblance measure, which may be either a linear correlation or distance measure. In Figure 3.35 there are different clustering results of two-dimensional inputs. Figure 3.35a is categorized according to distance. Such a method is referred to as the k-means in the literature (see Chapter 5). Provided there are two data series as X_i $(x_{i1}, x_{i2}, \ldots, x_{in})$ and X_j $(x_{j1}, x_{j2}, \ldots, x_{jn})$, the Euclidean distance, $d_{i,j}$, between two data points can be defined as:

$$d_{i,j} = \sqrt{\sum_{k=1}^{n} (x_{ik} - x_{jk})^2} \tag{3.26}$$

Another measure of resemblance is through the angle-difference (see Figure 3.35b) defined as:

$$\cos(X_i, X_j) = \frac{\sum_{k=1}^{n} x_{ik} x_{jk}}{\sqrt{\sum_{k=1}^{n} x_{ik}^2} \sqrt{\sum_{k=1}^{n} x_{jk}}} \tag{3.27}$$

This methodology assumes a linear dependence between points, and therefore must be used with caution. Of course, if there are m data sets, then there will be $m(m-1)/2$ distances or angle-differences among them with the application of Equations (3.26) and (3.27). As will be explained in Chapter 5, these methods yield the center of each category as shown by black circles in Figure 3.35.

After grouping the data by using some methodology into different categories, each cluster will have horizontal and vertical standard deviations in addition to

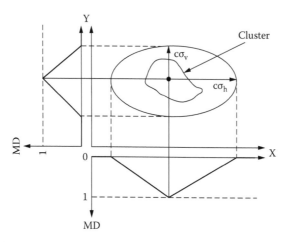

FIGURE 3.36 Cluster-based MF.

its center. The horizontal, σ_h, and vertical, σ_v, distances as standard deviations can be calculated for each category. After the center identification, in practice c times (1.5 or 2 times) the horizontal and vertical standard deviations represent each support, as shown in Figure 3.36. In this manner, for instance, a triangular MF is assumed to have MD equal to 1 at the central point along both axes, and the left and right boundaries on the horizontal and vertical axis c times the standard deviation. This procedure can be generalized and, in the case of more than one cluster overlapping MFs appear as in Figure 3.37. It is obvious that because

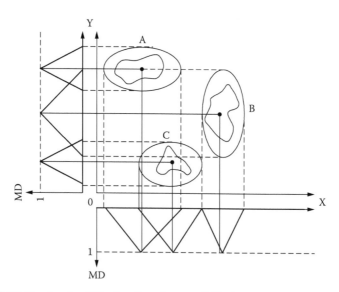

FIGURE 3.37 Overlapping MF allocations from available data.

there are three clusters, there will be, in general, three MFs for each variable. Sometimes an MF may fall inside another one. In this case, the outer MF must be adopted only (Chapter 7).

3.6 HEDGES (ADJECTIVIZED WORDS)

The linguistic variables are defined in terms of a single basic variable in CL context, but in real life quite often an adjective is attached in front of linguistic variables to express the ideas in a more refined manner. The very common linguistic hedges are "quite," "rather," "almost," "slightly," "approximately," "more or less," "very," etc. In general, the value of a linguistic variable is a composite term that is a concatenation of atomic words that can be classified into three parts:

1. The labels of fuzzy sets are the primary terms, such as *short, old, extensive,* etc.
2. Logical conjunctions, such as *not, and,* and *or.*
3. Hedges, such as *very, almost, more or less, very,* etc.

Much has been made about the relationship of FL to the human thought process and the ability to handle imprecise conditions that may arise. One of the terms frequently seen in the FL literature is the concept of *hedging,* which can be described as the modifiers to a certain set, much like the way adjectives and adverbs modify statements in the English language. When referring to a fuzzy set, hedges are used to adjust the characteristics of that fuzzy set by either one of the effects in Table 3.3.

A hedge is a one-input truth value manipulation operation. It modifies the shape of the MF, in a manner analogous to the function of adjectives and adverbs in English. Some examples that are commonly seen in the literature are intensifiers like "very," de-intensifiers like "somewhat," and complementary conjunctions like "not." One might define "very x" as the square of the truth value of x, and define "somewhat x" as the square root of the truth value of x. Then one can make fuzzy logic statements

TABLE 3.3
Hedges and Their Effects

Key Hedges	Effect on Set Characteristics
• "about"	
• "near"	Approximate the set
• "close to"	
• "approximately"	
• "not"	Complement the set
• "somewhat"	
• "rather"	Dilute the set
• "quite"	
• "very"	Intensify the set
• "extremely"	

such as "x is very 'low,'" which would evaluate this statement by multiplication as (x is "low")(x is "low"). One can think of "not x" as being a hedge in the same sense, defining "not x" as one minus the truth value of x.

FL systems have the ability to define hedges or modifiers of fuzzy values. In this manner, close ties are maintained to natural languages and language variables. These bring extra dimension to the fuzzy system flexibility and ultimate derivation of operations with the same formality as classic logic. The following are examples of hedges used on different types of variables:

1. General: "very," "quite," "rather," "extremely"
2. Truth value: "quite true," "almost true," "more or less true," "mostly true"
3. Probability: "likely," not very likely"
4. Quantity: "most," "several," "few"
5. Possibilistic: "almost impossible," "quite possible"

The simplest example is when one transforms the statement of "the flood is 'severe'" to "the flood is very 'severe.'" Some of the hedges may not have any numerical implication in everyday life but when used as hedges in the fuzzy context, their numerical implications become clear.

3.6.1   Fuzzy Reduction (Contraction)

The contraction operation, CON( ), reduces the fuzziness in a given fuzzy set. In the fuzzy domain if F is a fuzzy subset in U, then contraction is defined as a fuzzy set F in the same U set with the MF as:

$$\mu_{CON(F)}(x) = [\mu_F(x)]^2 \tag{3.28}$$

If the basic fuzzy set is adopted as a Gaussian MF (Equation (3.21)), then the normal form with $s = 2$ and $m = 0$ can be converted to "very" and "very very" by successive squaring according to Equation (3.28) and the results are presented in Figure 3.38. It can be observed that the MD attained by the concentration operation becomes lower as the original MD diminishes.

Hedges provide a natural and mathematically simple way of qualifying fuzzy variables. As a result, the expressive power of a fuzzy representation can be increased at little cost. Algorithmic operations can be devised that translate "fuzzy" terminology into numeric values, perform reliable operations upon those values, and then return natural language statements in a reliable manner.

3.6.2   Fuzzy Expansion (Dilatation)

On the other hand, the dilatation operation, DIL( ), of a fuzzy set F will have the following MF, by definition:

$$\mu_{DIL(F)}(x) = \sqrt{\mu_F(x)} \tag{3.29}$$

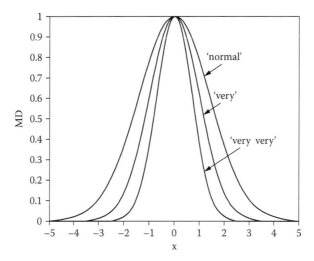

FIGURE 3.38 Contraction of MF.

The dilatation operation yields the results contrary to those achieved by the concentration operation. Hence, the MD increases as the original MD diminishes. The dilatation of the basic MF, and hence the fuzziness, increases as shown in Figure 3.39 for "slightly."

3.6.3 FUZZY REDUCTION-EXPANSION (INTENSIFICATION)

Another procedure in fuzzy numbers linguistic hedge computation is the intensification, INT( ), which results in a growing difference in the MD between elements

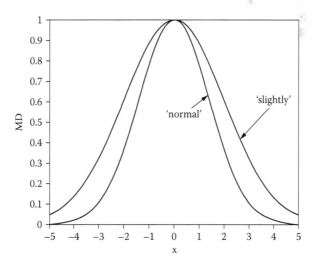

FIGURE 3.39 Dilatation of MF.

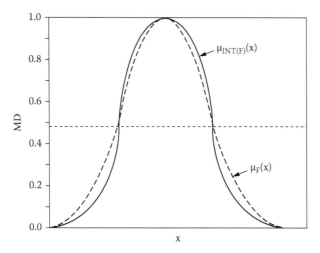

FIGURE 3.40 Intensification of MF.

whose original MD is greater than 0.5 and those whose original MD is less than 0.5. The intensification operation is achieved according to the following expression:

$$\mu_{INT(F)}(x) = \begin{cases} 2\big[\mu_F(x)\big]^2 & 0 < \mu_F(x) < 0.5 \\ 1 - 2\big[1 - \mu_F(x)\big]^2 & 0.5 < \mu_F(x) < 0 \end{cases} \qquad (3.30)$$

The intensification result is shown in Figure 3.40, again with a Gaussian MF.

3.7 LOGICAL OPERATIONS ON FUZZY SETS

The classical set theory defines four fundamental operations on sets: containment, complement ("NOTing"), intersection ("ANDing"), and union ("ORing"). These operations can be considered in relation to fuzzy sets. In addition, a large number of operations and manipulations can be applied to fuzzy sets that are not applicable to crisp sets. The basic linguistic operations on the two sets are equality, containment, complement, union, and intersection.

3.7.1 EQUIVALANCE

The equality says that two sets, A and B, are equal if and only if their MFs, $\mu_A(x) = \mu_B(x)$ for all the discourses over the universal set, U.

3.7.2 CONTAINMENT

A set can be completely contained in a larger set as its subset. Crisp subsets have memberships of 1 because their supersets are existing members of the specified

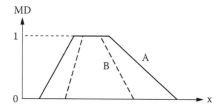

FIGURE 3.41 A contains B.

concept. The memberships of a fuzzy subset can have lower MDs than their super-sets. That is, an item can belong less to a subset than it does to its parent super-set. If $A = \{0/0 + 0.25/1 + 0.75/2 + 0/3\}$ and $B = \{0/0 + 0.15/1 + 0.45/2 + 0/3\}$, then it is possible to see that B is contained in A, which is represented symbolically as $\boldsymbol{B \subset A}$ if and only if the comparisons of corresponding MDs for the same discourse, x, in the universal set satisfies $\mu_A(x) \geq \mu_B(x)$. This is shown geometrically in Figure 3.41 with the positions of the MFs.

3.7.3 "ANDING" (INTERSECTION)

The intersection of two fuzzy subsets, A and B, is also a fuzzy set in U denoted by $A \wedge B$, where the $\wedge$ sign implies either in the linguistic sense logical "AND" conjunctive, which is shown symbolically for crisp sets as $A \cap B$, or numerically calculation of the MD for the intersection set from the basic sets. In the crisp sets, the consideration of A and B set elements as 0 and 1 leads to four possible combinations, as shown in the first two columns of Table 3.4.

The intersection in the last column is obtained by minimization of the MDs on each line. Hence, "ANDing" (i.e., intersection of two sets) requires a minimization operation for the final result. Similar operation is valid in the case of fuzzy subsets; therefore, the $\wedge$ sign in $A \wedge B$ implies fuzzy sets "ANDing," intersection or numerically the minimization of the MDs.

TABLE 3.4
A and B Crisp Set Intersection

	MD	
A	B	$A \cap B$
1	1	1
1	0	0
0	1	0
0	0	0

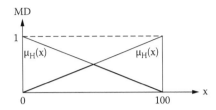

FIGURE 3.42 Intersection of two fuzzy subsets.

Hence, in notations, the intersection of these two fuzzy subsets can be calculated as follows:

$$\mu_{A \wedge B}(x) = \min[\mu_A, \mu_B] = \begin{cases} \mu_A(x) & \text{if} \quad 0 \le p(x) \le 0.5 \\ \mu_B(x) & \text{if} \quad 0.5 \le p(x) \le 1.0 \end{cases} \qquad (3.31)$$

which is illustrated in Figure 3.42.

Example 3.13

Find the intersection of the two fuzzy subsets "several" and "high" given in Table 3.5 and the operation of minimization that leads to the results shown in the rightmost column of this table.

The resulting combined fuzzy set of "several" ∧ "high" does not constitute a normal fuzzy set because it does not have at least one member with MD equal to 1. On the other hand, if two fuzzy sets are given as "long" = {0/0 + 0.25/1 + 0.75/2 + 1.0/3} and "short" = {1.0/0 + 0.75/1 + 0.25/2 + 0/3}, then the intersection can be obtained as:

$$\mu_{long}(x) \wedge \mu_{short}(x) = \{0/0 + 0.25/1 + 0.25/2 + 0/3\}$$

TABLE 3.5
Two Fuzzy Sets Union

Members	"several'	"high'	"several' ∧ "high"
		MDs	
3	0.4	0.0	0.4 ∧0.0 = 0.0
4	0.8	0.0	0.8 ∧0.0 = 0.0
5	1.0	0.2	1.0 ∧0.2 = 0.2
6	1.0	0.4	1.0 ∧0.4 = 0.4
7	0.8	0.6	0.8 ∧0.6 = 0.6
8	0.5	0.8	0.5 ∧0.8 = 0.5
9	0.0	1.0	0.0 ∧1.0 = 0.0

This is not a counterintuitive result. There are also different fuzzy sets' intersections proposed in the literature. For two fuzzy subsets A and B, the general form of t-norm can be written as:

$$t[\mu_A(x), \mu_B(x)] = \mu_{A \cap B}(x) \tag{3.32}$$

where $t[\ ]$ indicates the special functional form of any t-norms, which should have the following properties (Klir and Folger, 1998):

1. The boundary conditions should satisfy $t(0, 0) = 0$; $t(m_1, 1) = t(1, m_1) = m_1$.
2. The commutativity condition requires that $t(m_1, m_2) = t(m_2, m_1)$.
3. The associativity condition is $t(t(m_1, m_2), m_3) = t(m_1, t(m_2, m_3))$.
4. The non-decreasing should impose on the final fuzzy intersection set that if $m_1 < m_1'$ and $m_2 < m_2'$, then $t(m_1, m_2) \leq t(m_1', m_2')$.

The basic t-norm composition is already presented with the minimization operator as in the classical sets intersection, which reads $\min[\mu_A(x), \mu_B(x)] = \mu_{A \cap B}(x)$. Other types are given by Dubois and Parade (1980) and Yager (1980), respectively, as:

$$t_\lambda(m_1, m_2) = \frac{m_1 m_2}{\max(m_1, m_2, \lambda)} \quad (0 < \lambda < 1,) \tag{3.33}$$

and

$$t_\lambda(m_1, m_2) = 1 - \min\left\{1, - [1 - m_1]^\lambda + [1 - m_2]^{1/\lambda}\right\} \tag{3.34}$$

Some other t-norms without any parameter but depending on the MDs only are also available in the literature as the drastic product:

$$t_{dp}(m_1, m_2) = \begin{cases} m_1 & \text{if } m_2 = 1 \\ m_2 & \text{if } m_1 = 1 \\ 0 & \text{otherwise} \end{cases} \tag{3.35}$$

or as the Einstein product:

$$t(m_1, m_2) = \frac{m_1 m_2}{2 - (m_1 + m_2 - m_1 m_2)} \tag{3.36}$$

and finally as the algebraic product:

$$t(m_1, m_2) = m_1 m_2 \tag{3.37}$$

TABLE 3.6

A and B Crisp Set Union

MD		
A	*B*	*A* ∪ *B*
1	1	1
1	0	1
0	1	1
0	0	0

3.7.4   "ORING" (UNION)

The union of two fuzzy subsets, A and B, is also a fuzzy set in U denoted by $A \vee B$, where the $\vee$ sign implies, in the linguistic sense, the logical "OR" conjunction. In the crisp sets, the union is shown as $A \cup B$, which leads to four possible combinations as shown in the last column of Table 3.6. The union in the last column is obtained by maximization of the MDs on each line. A similar operation is valid in the case of fuzzy subsets; therefore, the $\vee$ sign in $A \vee B$ implies fuzzy sets "ORing," union, or numerically the maximization of the MDs.

Example 3.14

If two fuzzy sets "several" and "high" have the following elements with the corresponding MDs, find the union of the two sets:

$$\text{'several'} = \left\{ \frac{0.4}{3} + \frac{0.8}{4} + \frac{1.0}{5} + \frac{1.0}{6} + \frac{0.8}{7} + \frac{0.5}{8} \right\}$$

and

$$\text{'high'} = \left\{ \frac{0.2}{5} + \frac{0.4}{6} + \frac{0.6}{7} + \frac{0.8}{8} + \frac{1.0}{9} \right\}$$

A similar table can be prepared (Table 3.7) that includes the member values in the first column, their corresponding MDs in the second and third columns, and the final column is for the results of "ORing," union, or maximization of the "several" and "high" fuzzy sets.

It must be noted that member "3" is not included in the "high" fuzzy set. It means that its MD of belonging to this fuzzy subset is equal to 0. The same argument is valid for member "9," which is a non-member for the "several" fuzzy subset.

The union of $\mu_A(x)$ and $\mu_B(x)$ can be written mathematically similar to Equation (3.31) as:

$$m_{A \vee B}(x) = \max[\mu_A, \mu_B] = \begin{cases} \mu_A(x) & \text{if } 0 \leq p(x) \leq 0.5 \\ \mu_B(x) & \text{if } 0.5 \leq p(x) \leq 1.0 \end{cases} \tag{3.38}$$

and is plotted in Figure 3.43.

TABLE 3.7

Two Fuzzy Sets Union

	MDs		
Members	"several"	"high"	"several" ∨ "high"
3	0.4	0.0	0.4 ∨ 0.0 = 0.4
4	0.8	0.0	0.8 ∨ 0.0 = 0.8
5	1.0	0.2	1.0 ∨ 0.2 = 1.0
6	1.0	0.4	1.0 ∨ 0.4 = 1.0
7	0.8	0.6	0.8 ∨ 0.6 = 0.8
8	0.5	0.8	0.5 ∨ 0.8 = 0.8
9	0.0	1.0	0.0 ∨ 1.0 = 1.0

Following similar arguments to the "ANDing," the "ORing" of two given sets "long" and "short" as in the previous section leads to:

$$\mu_{long}(x) \vee \mu_{short}(x) = \{1.0/0 + 0.75/1 + 0.75/2 + 1.0/3\}$$

Again, this is not a counterintuitive result. In the ordinary union of two sets, as explained in the previous section, two sets are mapped into a single set with MFs. It can also be shown in the form of a functional transformation in general as:

$$s[\mu_A(x), \mu_B(x)] = \mu_{A \vee B}(x) \tag{3.39}$$

where $s[\ ]$ indicates the special form of any s-norms. The simplest union is presented for crisp sets similar to Equation (3.39) as $\max[m_A(x), m_B(x)] = \mu_{A \cup B}(x)$. The question now is whether it is possible to find s functions that qualify as a union. In general, s-norms should have the following properties (Klir and Folger, 1988):

1. The MDs must vary between 0 and 1, inclusively. This can be regarded as a boundary condition. According to Table 3.6, one can write $s(1, 1) = 1$, $s(0, m) = s(m, 0) = m$.
2. A sort of symmetry (commutative condition) must also be satisfied as $s(m_1, m_2) = s(m_2, m_1)$.

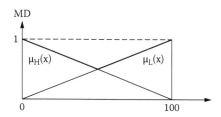

FIGURE 3.43 Union of two fuzzy subsets.

3. The associative property must be valid also for union fuzzy sets as $s(s(m_1, m_2), m_3) = s(m_1, s(m_2, m_3))$,
4. Of course, the fuzzy set would also have a non-decreasing condition imposed on it as if $m_1 < m_1'$ and $m_2 < m_2'$, then $s(m_1, m_2) \leq s(m_1', m_2')$.

These four conditions can be proven with the classical union formulation. However, there are a few other s-norms of fuzzy union calculation suggested by Dubois and Prade (1980) and Yager (1980), respectively, as:

$$s_\lambda(m_1, m_2) = \frac{m_1 + m_2 - m_1 m_2 - \min(m_1, m_2, 1 - \lambda)}{\max(1 - m_1, 1 - m_2, \lambda)} \tag{3.40}$$

and

$$s_\lambda(m_1, m_2) = \min\left[1, \left(m_1^\lambda + m_2^\lambda\right)^{1/\lambda}\right] \tag{3.41}$$

With the suitable choice of the λ parameter, the union of many s-norm fuzzy subsets can be obtained. Some of the other s-norm formulations are in different summation forms, such as the drastic sum:

$$s_{ds}(m_1, m_2) = \begin{cases} m_1 & \text{if } m_2 = 0 \\ m_2 & \text{if } m_1 = 0 \\ 0 & \text{otherwise} \end{cases} \tag{3.42}$$

or as Einstein sum:

$$s(m_1, m_2) = \frac{m_1 + m_2}{1 + m_1 m_2} \tag{3.43}$$

The algebraic sum gives the following expression:

$$s(m_1, m_2) = m_1 + m_2 - m_1 m_2 \tag{3.44}$$

3.7.5 "NOTing" (Complement)

Given a fuzzy set A, its complement $\bar{A}$ is defined by its MF simple as:

$$\mu_{\bar{A}}(x) = 1 - \mu_A(x) \tag{3.45}$$

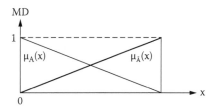

FIGURE 3.44 Union of two fuzzy subsets.

For example, if $A = \{0/0 + 0.25/1 + 1.0/2 + 0/3\}$, then $\overline{A} = \{1/0 + 0.75/1 + 0.00/2 + 1/3\}$. Figure 3.44 shows the "NOTing" operation for the X linguistic variable, which has its support from 0 to 100.

The requirement for complement is to transform the fuzzy set A into another one as:

$$\mu_{\overline{A}}(x) = f[\mu_A(x)] \tag{3.46}$$

The problem is the definition of the transformer $f(.)$. From the crisp sets theory, its simplest form appears as in Equation (3.45). Although this is the only way of obtaining the complement for classical sets, in the case of fuzzy subsets, there are different types of complements. This is due to the fact that the MDs assume not only 0 and 1 as in the crisp sets, but any value between 1 and 0, inclusively. The necessary requirements for a set to be the complement of another is that (Klir and Folger, 1988):

1. Its MDs should vary between 0 and 1.
2. If the MDs m_1 and m_2 of two members, then $f(m_1) \geq f(m_2)$, which implies a non-decreasing condition.

Several functions satisfying these conditions have been proposed by Sugeno (1977) as:

$$f(x) = \frac{1 - \mu_A(x)}{1 + \lambda\mu_A(x)} \tag{3.47}$$

where $-1 < \lambda < \infty$. For $\lambda = 0$, the simplest complement appears as in Equation (3.45). Sugeno complements are indicated in Figure 3.45 for various λ values.

Another type of fuzzy complement was given by Yager (1980) and defined by the following complement equation:

$$f(x) = \left[1 - \mu_A^{\lambda}(x)\right]^{1/\lambda} \tag{3.48}$$

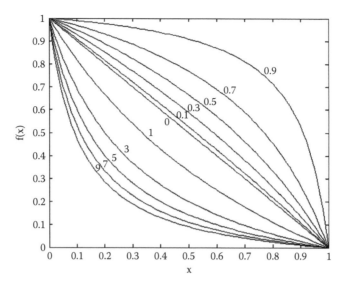

FIGURE 3.45 Sugeno complement function.

This expression reduces to Equation (3.45) for $\lambda = 1$. The graphical presentation is given in Figure 3.46.

3.7.6 DE MORGAN'S LAW

De Morgan's laws can be written for the fuzzy sets of A and B, similarly to crisp sets as:

$$\overline{A \vee B} = \overline{A} \wedge \overline{B} \tag{3.49}$$

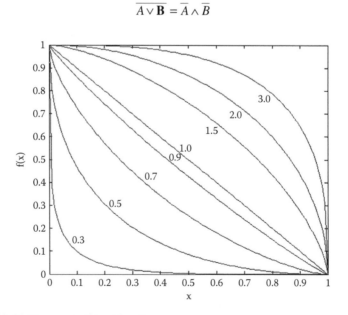

FIGURE 3.46 Yager compliment function.

and vice versa:

$$\overline{A \wedge B} = \overline{A} \vee \overline{B} \qquad (3.50)$$

where overbars indicate the complement of the fuzzy set concerned. These can be proved by first showing that the following identity is true:

$$1 - \max[\mu_A(x), \mu_B(x)] = \min[1 - \mu_A(x), 1 - \mu_B(x)]$$

The two possible cases are:

$$\mu_A(x) \geq \mu_B(x)$$

and

$$\mu_A(x) \leq \mu_B(x).$$

If $\mu_A(x) \geq \mu_B(x)$ then $1 - \mu_A(x) \leq 1 - \mu_B(x)$
and

$$1 - \max[\mu_A(x), \mu_B(x)] = \min[1 - \mu_A(x), 1 - \mu_B(x)]$$

On the other hand, if $\mu_A(x) \leq \mu_B(x)$, then $1 - \mu_A(x) > 1 - \mu_B(x)$ and

$$1 - \max[\mu_A(x), \mu_B(x)] = 1 - \mu_B = \min[1 - \mu_A(x), 1 - \mu_B(x)]$$

which corresponds to Equation (3.31).

Apart from the above classical complement, union, and intersection operations, there are other alternatives for each one of these operations that do not exist in the classical sets (Wang, 1997).

The Law of the Excluded Middle and the Law of Contradiction of the Crisp Sets are no longer true and valid in fuzzy sets; that is, for fuzzy set A:

$$A \vee \overline{A} = U \qquad (3.51)$$

and

$$A \wedge \overline{A} = \Phi \qquad (3.52)$$

where U and Φ are the universal and empty sets, respectively. Equations (3.51) and (3.52) mean that due to a lack of precision (crisp) boundaries, complementary sets are overlapping and cannot cover the universal set completely. On the contrary, these two laws are the necessary characteristics of crisp sets and CL.

3.7.7 FUZZY AVERAGING

It can be understood from previous discussions that in the case of two fuzzy sets A and B, their MDs after the union and intersection will lie between the ranges of [max (m_1, m_2), s-norm(m_1, m_2)] and [t-norm(m_1, m_2), min (m_1, m_2)], respectively. This means that both operators (union and intersection) cannot cover the interval between min (m_1, m_2) and max (m_1, m_2). To cover this interval, there are fuzzy averaging operators that convert two fuzzy subsets into a single one with different operators, as given in the literature (Wang, 1997; Ross, 1995). Some of the averaging operators are given in the following as the maximum-minimum average (Klir and Folger, 1988):

$$a_\lambda(m_1, m_2) = \lambda \max(m_1, m_2) + (1-\lambda)\min(m_1, m_2) \quad (0 \le \lambda \le 1) \tag{3.53}$$

or as the generalized average:

$$a_\lambda(m_1, m_2) = \left(\frac{m_1^\lambda + m_2^\lambda}{2}\right)^{1/\lambda} \quad (\lambda \ne 0) \tag{3.54}$$

or as fuzzy "ANDing":

$$a_\lambda(m_1, m_2) = \lambda \min(m_1, m_2) + \frac{(1-\lambda)(m_1 + m_2)}{2} \quad (0 \le \lambda \le 1) \tag{3.55}$$

The fuzzy "ORing" is given as

$$a_\lambda(m_1, m_2) = \lambda \max(m_1, m_2) + \frac{(1-\lambda)(m_1 + m_2)}{2} \quad (0 \le \lambda \le 1) \tag{3.56}$$

It is obvious that the max-min operator covers the whole interval [min (m_1, m_2), max (m_1, m_2)] as the parameter λ takes values from 0 to 1. On the other hand, the fuzzy "ORing" covers the interval from min (m_1, m_2) to $(m_1 + m_2)/2$, and the fuzzy "ANDing" from $(m_1 + m_2)/2$ to max (m_1, m_2).

REFERENCES

Benjamin, J.R. and Cornell, C.A. 1973. *Probability Statistics and Decision Making in Civil Engineering*. McGraw-Hill, New York.

Dubois, D. and Prade, H. 1980. *Fuzzy Sets and Systems. Theory and Applications*. Academic Press, Orlando, FL.

Klir, G.J. and Folger, T.A. 1988. *Fuzzy Sets, Uncertainty and Information*. Prentice-Hall, Englewood Cliffs, NJ.

Löf, G.O.G, Duffie, J.A., Smith, C.O. 1966. World distribution of solar radiation. *Solar Energy*, 10: 27–37.

Mendel, J.M. 1995. Fuzzy logic system for engineering: a tutorial. *Proc. IEEE* 83: 345–377.

Ross, J.T. 1995. *Fuzzy Logic with Engineering Applications*. McGraw-Hill, New York, 593 pp.

Sugeno, M. 1977. Fuzzy measures and fuzzy integrals: a survey. In: Gupta, M., G.N. Saridis, and B.R. Gaines, Eds., *Fuzzy Automata and Decision Processes,* North-Holland, New York, pp. 329–346.

Şen, Z. 2004. *Genetik Algoritmalar ve Eniyileme Yöntemleri* (Genetic Algorithms and Optimization Methods). Su Vakfı Yayınları, 142 pp. (in Turkish).

Wang, L.X. 1997. *A Course in Fuzzy Systems and Control.* Prentice-Hall, Englewood Cliffs, NJ, 424 pp.

Yager, R.R. 1980. On a general class of fuzzy connectives. *Fuzzy Sets and Systems* 4(3); 253–242.

Zadeh, L.A. 1965. Fuzzy sets. *Informat. and Control* 8: 338–353.

PROBLEMS

3.1 Give a mutually exclusive sub-range for each of the following words from the universe between 0 and 100:
   (a) "fair"
   (b) "good"
   (c) "excellent"
   (d) "weak"

3.2 Give a mutually inclusive sub-range for each of the following words from the universe between 0 and 100:
   (a) "excellent"
   (b) "weak"
   (c) "good"
   (d) "fair"

3.3 Develop a reasonable sub-range for each of the following words based on "porosity":
   (a) "silt"
   (b) "sand"
   (c) "gravel"
   (d) "clay"

3.4 Provide a domain of change for each one of the following words on the basis of "rainfall":
   (a) "torrential"
   (b) "intensive"
   (c) "medium"
   (d) "low"

3.5 Identify the approximate range of numbers implied by the following words:
   (a) "few"
   (b) "several"
   (c) "a lot"
   (d) "crowd"

3.6 Indicate the following words on a Cartesian coordinate system:
   (a) X ("cold," "cool," "warm")
   (b) Y ("wet," "medium," "dry")

3.7 Which numbers would you attach to the following water quality words from 100 to 1000?
   (a) "brackish"
   (b) "saline"
   (c) "potable"
   (d) "fresh"

3.8 Disintegrate "discharge" into at least three categories in descending order from the following words:
(a) "low"
(b) "intense"
(c) "high"
(d) "small"
(e) "weak"
(f) "dry"

3.9 Discuss whether the cumulative probability distribution (CDF) in statistics can be considered a fuzzy set and membership function.

3.10 If the groundwater potentiality P has a crisp set of values as $P = \{0, 5, 50, 500, 1000\}$, convert this set into a fuzzy set according to the following criteria:
(a) "low potential"
(b) "moderate potential"

3.11 The relative frequency diagram of a rainfall event in an area has the values given in the table. Is it possible to convert it into a fuzzy set?

Class Interval	1–3	3–5	5–7	7–9	9–11	11–13	13–15	15–17
	0.01	0.02	0.15	0.20	0.32	0.17	0.11	0.01

3.12 In the following graphs, rainfall and runoff variables have in their indicated domain of variation three fuzzy sets each. Are these acceptable? If there are defects, correct accordingly.

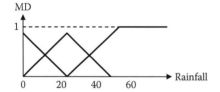 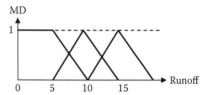

3.13 If the drought fuzzy description is presented by three fuzzy sets ("dry," "medium," and "wet") as in the following figure, what are the MDs of the following runoff R values in [m³/sec]. $R = \{3.2, 134.5, 35.7, 89.1, 321.5, 214.3, 156.3\}$

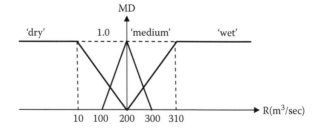

3.14 Use your intuition and develop the following MFs for peak discharge of 210 m³/sec:
(a) Sigmoid
(b) Symmetric triangle
(c) Trapezoid

3.15 Use your intuition and develop the following MFs for "approximately 10 mm/day" OR "approximately 21 mm/day" rainfall.
  (a) Gaussian functions
  (b) Z- shapes
  (c) Symmetric triangles

3.16 Define the universe of discourse and plot MFs, using your intuition about the following cases:
  (a) Infiltration ("very slow," "slow," "moderate," high," "very high")
  (b) Evaporation rate ("very low," "low," "medium," "big," "very big")
  (c) Water quality ("saline," "brackish," "moderate," "potable," "pure")

3.17 In an area, the groundwater potentiality is "high," with 450 m²/day transmissivity and the water quality is "good" at 500 ppm. Plot the fuzzy set using symmetric MFs for the following alternatives:
  (a) "high" potentiality and "good" quality
  (b) "high" potentiality or "good" quality
  (c) not "high" potentiality but "good" quality

3.18 Perform an "ANDing" operation between A and B and then an "ORing" operation between the result and set C.

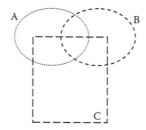

3.19 Which one of the following graphs has fuzzy features?

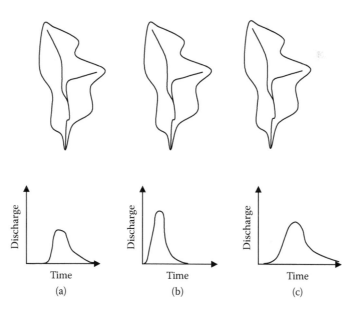

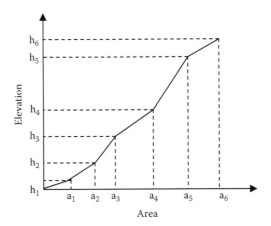

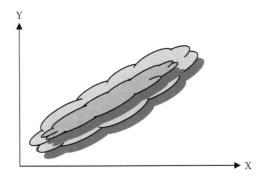

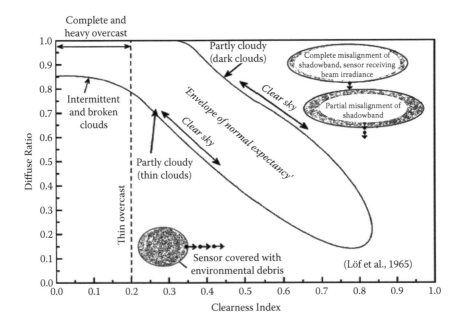

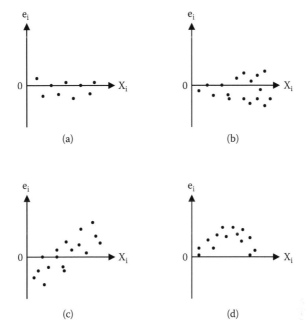

3.20 Explain the difference between the following drainage basin classifications on the basis of surface area classifications.

(a)

Area (km²)	Classification
A >1000	"very big"
1000 < A < 100	"big"
100 < A < 5	"middle"
A < 5	"small"

(b)

Area (km²)	Classification
A >1000	"very big"
1200 < A < 80	"big"
80 < A < 3	"middle"
A < 5	"small"

3.21 Do the following fuzzy sets satisfy De Morgan's law?

(a) $\mu_A(x) = \dfrac{1}{1+2|x|}$

(b) $\mu_A(x) = \sqrt{\dfrac{1}{1+2|x|}}$

3.22 Express the following statements according to appropriate logic, and draw their MFs:
(a) Real numbers greater than 7
(b) Real numbers close to 7

3.23 If two experts (E_1 and E_2) have stated that the transmissivity T of an aquifer has "high" potential with the following fuzzy sets, then what is the common decision that comes out of these two independent suggestions?

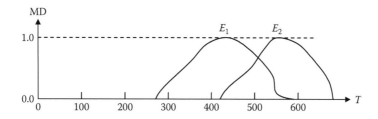

4 Fuzzy Numbers and Arithmetic

4.1 GENERAL

Recent developments in the operations and computations of fuzzy numbers evolved into fuzzy arithmetic, which has stimulated researchers and engineers alike to delve into the benefits of possibility theory rather than probability theory. Fuzzy numbers are everywhere in daily life, and almost all assessments without modeling are expressed with approximate numbers. It is therefore necessary to know how to deal with their arithmetic operations in order to combine individual statements in a combination depending on requirements, whether as addition, subtraction, multiplication, or division. Fuzzy numbers are essential for expressing fuzzy quantities, and fuzzy arithmetic is a basic tool in dealing with fuzzy quantifiers in approximate reasoning (see Chapter 2). Fuzzy arithmetic is the generalization of interval arithmetic, which is used in various disciplines for dealing with the inaccuracies of measuring instruments in the form of intervals. Such intervals are based on measurements such as the precision or confidence intervals in statistical research. Hence, fuzzy arithmetic has more far-reaching expressive power than interval arithmetic. Fuzzy arithmetic considers intervals or several levels between 0 and 1 inclusive, whereas interval arithmetic has CL with a unique level over the whole support (variation domain).

Kaufmann and Gupta (1985) wrote a complete book on the fuzzy arithmetic where different arithmetical and mathematical operations on fuzzy sets are presented. They have noted that fuzzy numbers can be treated as a generalization of the concept of the confidence interval.

In this chapter, fuzzy arithmetic operations are explained with examples. Classes with unsharp boundaries are the basis of fuzzy sets. Their fundamental principles, concepts, and mathematical presentations are given in this chapter with emphasis on water-related phenomena.

4.2 FUZZY NUMBERS

Fuzzy numbers are normal and convex fuzzy sets along the support line. There are certain rules for the arithmetic and mathematical operations on the fuzzy sets, which are rather different from the ordinary sets or numbers, and are collectively called fuzzy arithmetic operations. In the special case where the same MDs are all equal to 1 or 0, these operations are identical with the commonly known arithmetic operations.

Subjective estimations in hydrology are very common and, depending on intuition and previous experience, a hydrologist can state that pump discharge is "around

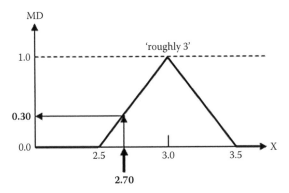

FIGURE 4.1 Fuzzy number 3.

3 L/sec" or that the cost of per-meter drilling is "approximately 100 USD." These subjective estimations are characterized by certain numerical values, which are comparable. For instance, the discharge estimate "around 3 L/sec" is more than a discharge estimation of "around 2 L/sec" Numbers expressed in this manner are called fuzzy numbers. A fuzzy number is simply an ordinary number whose precise value is somewhat uncertain, and is represented by a fuzzy set that is convex and normal. For example, Figure 4.1 presents a fuzzy 3. The MDs for numbers 2.5 or less and 3.5 or greater than fuzzy 3 are all equal to 0, but between these two boundaries all the numbers are attributed to fuzzy 3 with various degrees, including "exactly 3" core) with the MD equal to 1. For instance, the MD of 2.7 belonging to the fuzzy 3 is 0.30. For numbers greater than 3, the MD declines from 1 to 0 as the support value increases. A very convenient way to describe fuzzy numbers is to use modifying words (hedges, adjectives). For example, the fuzzy number 3 shown in Figure 4.1 could be specified completely by the statement "roughly 3." Other available modifying words may be *nearly, about, approximately, around,* and *crudely.* It is important to notice that these words imply progressively larger uncertainties. They are called hedges in fuzzy logic terminology (see Section 3.6). It is possible to make approximate comparisons with fuzzy numbers. For example, to ask if the rainfall amount is "approximately equal to 30 cm" implies a fuzzy number. This is often very useful when data are imprecise, or when one does not want to accept rigidity that the rainfall amount is "exactly 30 cm." It is also equivalent to accepting that there are some uncertainty bands "around 30 cm" rainfall, which drops gradually on two sides with increasing uncertainty ingredient.

In approximate velocity measurements of a stream, hydrologists try to measure the time, t, between two fixed cross sections along a channel alignment. One can either make an initial guess as saying that it is "approximately t_0" or it cannot be "not less than t_1" and "not greater than t_2." This is tantamount to saying that the travel time is confined in the interval $[t_1, t_2]$. This is referred to as the confidence interval, which can be denoted as $T = [t_1, t_2]$. If this definition is adopted for Figure 4.1, then the confidence interval corresponds to $T = [2.5, 3.5]$. Here,

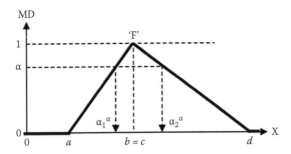

FIGURE 4.2 Confidence interval levels.

another definition attachment of the classical confidence interval appears, such that it is defined at zero MD level. Let us define the level by α, hence the confidence interval level corresponding to zero MD can be written as $\alpha = 0$. It is obvious that $0 < \alpha < 1$, and the confidence interval in Figure 4.1 at $\alpha = 1$ level has $T = [3, 3]$; that is, it is an ordinary number singleton (Chapter 3). As the α level increases, the confidence interval becomes narrower. Figure 4.2 indicates fuzzy number F for which the confidence interval corresponding to the α level is denoted as $[\alpha_1{}^\alpha, \alpha_2{}^\alpha]$.

Any MF mentioned in Chapter 3 can be adopted as a fuzzy number. For instance, trapezium fuzzy numbers are special cases of trapezium numbers, as shown in Figure 4.3 with four ordinary numbers, a, b, c, and d, where a and d are the left and right boundaries at $\alpha = 0$ cut-off level and intermediate values, namely, b and c are the left and right boundaries at $\alpha = 1$ cut-off level.

Fuzzy number arithmetic operations are rather different from ordinary numbers because they have lower and upper boundaries with the ends of confidence intervals at different levels. In four basic arithmetic operations, fuzzy number operations can be understood provided that the same operations are carried out on the basis of interval numbers, which are simpler to grasp.

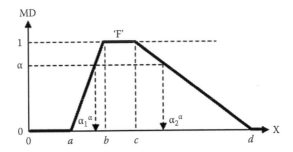

FIGURE 4.3 Trapezium fuzzy number.

Example 4.1

A fuzzy number "A" is given with its discrete elements and MDs as follows:

$$"A" = \left\{ \frac{0.3}{2} + \frac{0.5}{3} + \frac{0.8}{4} + \frac{1.0}{5} \right\}$$

Find the crisp sets at $\alpha = 0.3, 0.5, 0.8$, and 1.0 levels. The answers are as follows:

$$A_{0.3} = \{2,3,4,5\}$$

$$A_{0.5} = \{3,4,5\}$$

$$A_{0.8} = \{4,5\}$$

$$A_{1.0} = \{5\}$$

The first three α-cuts sets represent interval numbers but the last one is equivalent to an ordinary number, 5. To construct the same fuzzy number from its crisp sets ($\alpha = 0$), it is necessary to consider the following steps:

1. Write the crisp sets in the form of fuzzy sets but with MDs equal to 1.
2. Multiply each crisp set with the corresponding α-cut level.
3. Use "ORing" operation among the fuzzy sets at the last column for the final product.

All these steps are shown in Table 4.1.

TABLE 4.1
Cuttings

α	A_α (Crisp Sets)	αA_α (Fuzzy Sets)
0.3	$A = \left\{ \dfrac{1}{2} + \dfrac{1}{3} + \dfrac{1}{4} + \dfrac{1}{5} \right\}$	$A = \left\{ \dfrac{0.3}{2} + \dfrac{0.3}{3} + \dfrac{0.3}{4} + \dfrac{0.3}{5} \right\}$
0.5	$A = \left\{ \dfrac{1}{3} + \dfrac{1}{4} + \dfrac{1}{5} \right\}$	$A = \left\{ \dfrac{0.5}{3} + \dfrac{0.5}{4} + \dfrac{0.5}{5} \right\}$
0.8	$A = \left\{ \dfrac{1}{4} + \dfrac{1}{5} \right\}$	$A = \left\{ \dfrac{0.8}{4} + \dfrac{0.8}{5} \right\}$
1.0	$A = \left\{ \dfrac{1}{5} \right\}$	$A = \left\{ \dfrac{1}{5} \right\}$
	"ORing"	$A = \left\{ \dfrac{0.3}{2} + \dfrac{0.5}{3} + \dfrac{0.8}{4} + \dfrac{1.0}{5} \right\}$

4.3 FUZZY ADDITION

To understand the fuzzy number summation, it is better to consider first the confidence intervals addition. The support ($\alpha = 0$ cut) of any fuzzy is equal to the interval number. If two interval numbers are "X" = $[x_1, x_2]$ and "Y" = $[y_1, y_2]$, then their summation becomes:

$$\text{"}Z\text{"} = \text{"}X\text{"} + \text{"}Y\text{"} = [x_1 + y_1, x_2 + y_2] \tag{4.1}$$

The left and right boundaries of the sum of the confidence intervals are, respectively, equal to the sum of the left boundaries and the sum of the right boundaries of the confidence intervals being added. Similarly, the fuzzy numbers can be added by considering confidence intervals at α cut-off level. The two fuzzy numbers are added at the same α-cut level. Hence, the addition of two fuzzy numbers "X" and "Y" results in a new fuzzy number "Z" as:

$$\text{"}Z\text{"} = \text{"}X\text{"} + \text{"}Y\text{"} = [x_1{}^\alpha + y_1{}^\alpha, x_2{}^\alpha + y_2{}^\alpha] \tag{4.2}$$

Example 4.2

Let two small rivers, R_1 and R_2, confluence at point A as shown in Figure 4.4 with their subjective discharge estimates as follows: "Q_1" is "approximately 15 m³/sec" and "Q_2" is "approximately 5 m³/sec." Find the discharge "Q" after the confluence point.

In Figure 4.4, at α = 0 level, the confidence intervals (support) of fuzzy numbers are "Q_1" = [0, 10] and "Q_2" = [10, 20]. Triangular numbers are most often encountered in many applications and, in general, fuzzy numbers do not have triangular shapes, but let us assume that numbers "Q_1" and "Q_2" are represented by triangular MFs. The summation of two fuzzy number discharges will also result in another fuzzy discharge number "Q," which can be shown symbolically as:

$$\text{"}Q\text{"} = \text{"}Q_1\text{"} + \text{"}Q_2\text{"} \tag{4.3}$$

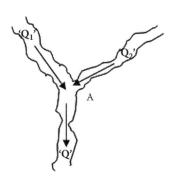

FIGURE 4.4 Confluence point.

To establish the MF of the resulting fuzzy discharge, let us write the fuzzy numbers MFs for "Q_1" and "Q_2" as:

$$\mu_{Q_1}(x) = \begin{cases} 0 & x < 0 \\ \dfrac{x}{5} & 0 < x < 5 \\ -\dfrac{x}{5} + 2 & 5 < x < 10 \\ 0 & x > 10 \end{cases}$$

(4.4)

and

$$\mu_{Q_2}(x) = \begin{cases} 0 & x < 10 \\ \dfrac{x}{5} - 2 & 10 < x < 15 \\ -\dfrac{x}{5} + 4 & 15 < x < 20 \\ 0 & x > 20 \end{cases}$$

(4.5)

An α-cut level yields the left and right boundaries of the corresponding confidence interval in each fuzzy number. For instance, the left and right boundaries of fuzzy number "Q_1" at α level are, respectively:

$$\alpha = \left(\frac{q_{1_1}^{\alpha}}{5} \right)$$

(4.6)

and

$$\alpha = -\left(\frac{q_{1_2}^{\alpha}}{5} \right) + 2$$

(4.7)

Hence, the left and right boundaries of the confidence interval at α level become:

$$q_{1_1}^{\alpha} = 5\alpha$$

(4.8)

and

$$q_{1_2}^{\alpha} = -5\alpha + 2$$

(4.9)

This means that the confidence interval $Q_{1\alpha}$ of the fuzzy number "Q_1" at the level of α is $Q_{1\alpha} = [5\alpha, -5\alpha + 10]$. Likewise, the confidence interval of fuzzy number "Q_2" becomes $Q_{2\alpha} = [5\alpha + 10, -5\alpha + 20]$. The summation fuzzy number will have confidence intervals according to Equation (4.2) as $Q_{\alpha} = [10\alpha + 10, -10\alpha + 30]$.

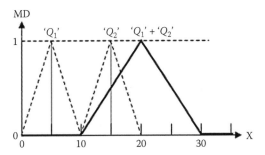

FIGURE 4.5 Confidence intervals.

This gives the left- and right-hand side boundaries, respectively, of the resulting fuzzy number as:

$$\alpha = \frac{q^\alpha}{10} - 1 \tag{4.10}$$

and

$$\alpha = -\frac{q^\alpha}{10} + 3 \tag{4.11}$$

The final form of the desired fuzzy total discharge value can be written as:

$$\mu_Q(x) = \begin{cases} 0 & x < 10 \\ \dfrac{x}{10} - 1 & 10 < x < 20 \\ -\dfrac{x}{10} + 3 & 20 < x < 30 \\ 0 & x > 30 \end{cases} \tag{4.12}$$

Hence, the final fuzzy number is shown in Figure 4.5.

4.4 FUZZY SUBTRACTION

Given the same "X" and "Y" confidence intervals as in the previous section, the resulting subtraction of "Y" from "X" appears as a new confidence interval "Z":

$$\text{“}Z\text{”} = \text{“}X\text{”} - \text{“}Y\text{”} = [x_1 - y_2, x_2 - y_1] \tag{4.13}$$

The left (right) boundary is the difference between the right (left) boundary of the left (right) boundary of the first confidence interval number and the right (left) boundary of the second confidence interval. Accordingly, the subtraction of two fuzzy numbers ("X" and "Y") leads to a new fuzzy number at α cut-off level as:

$$\text{“}Z\text{”} = \text{“}X\text{”} - \text{“}Y\text{”} = [x_1^\alpha - y_2^\alpha, x_2^\alpha - y_1^\alpha] \tag{4.14}$$

Example 4.3

The rainfall volume over a catchment area is "about 17×10^6 m³/year," which gives rise to "approximately 9×10^6 m³/year" surface runoff. Find the approximate total loss during the same year in this catchment. The "approximate rainfall" "R" and "approximate runoff" "r" volumes, respectively, are given again by triangular MFs as:

$$\mu_R(x) = \begin{cases} 0 & x < 14 \\ \dfrac{x}{3} - \dfrac{14}{3} & 14 < x < 17 \\ -\dfrac{x}{3} + \dfrac{20}{3} & 17 < x < 20 \\ 0 & x > 20 \end{cases} \tag{4.15}$$

and

$$\mu_r(x) = \begin{cases} 0 & x < 8 \\ x - 8 & 8 < x < 9 \\ -x + 10 & 9 < x < 10 \\ 0 & x > 10 \end{cases} \tag{4.16}$$

The confidence interval values at α cut-off level for "R" and "r" fuzzy numbers are "R" = $[3\alpha + 14, -3\alpha + 20]$ and "r" = $[\alpha + 8, -\alpha + 10]$, respectively. Hence, at the same cut-off level, the confidence interval of the resultant loss "L" fuzzy number by considering Equation (4.14) becomes "L" = $[2\alpha + 6, -2\alpha + 10]$. Accordingly, the left- and right-hand side boundaries of fuzzy number "L" are, respectively:

$$l_1^\alpha = 2\alpha + 6 \tag{4.17}$$

and

$$l_{21}^\alpha = -2\alpha + 10 \tag{4.18}$$

The equations of the lines representing the left and right boundaries of fuzzy number "L" are as follows:

$$\alpha = \left(\frac{l_1^\alpha}{2}\right) - 3$$

and

$$\alpha = -\left(\frac{l_1^\alpha}{2}\right) + 5$$

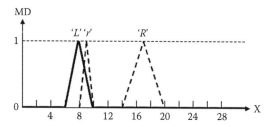

FIGURE 4.6 Fuzzy number subtraction.

It is then possible to write the loss fuzzy number as:

$$\mu_L(x) = \begin{cases} 0 & x < 6 \\ \dfrac{x}{2} - 3 & 6 < x < 8 \\ -\dfrac{x}{2} + 5 & 8 < x < 10 \\ 0 & x > 10 \end{cases} \qquad (4.19)$$

All the fuzzy numbers are shown collectively in Figure 4.6.

4.5 FUZZY MULTIPLICATION

Multiplication of two confidence interval numbers "X" and "Y" leads to a third confidence interval number "Z":

$$\text{``}Z\text{''} = \text{``}X\text{''}.\text{``}Y\text{''} = [\, x_1 y_1,\ x_2 y_2 \,] \qquad (4.20)$$

The left (right) boundary is equal to the multiplication of the left (right) boundary of the first confidence interval by the right (left) boundary of the second confidence interval. Likewise, the multiplication of two fuzzy numbers will be a new "Z" fuzzy number:

$$\text{``}Z\text{''} = \text{``}X\text{''}.\text{``}Y\text{''} = [x_1{}^\alpha y_2{}^\alpha,\ x_2{}^\alpha y_1{}^\alpha] \qquad (4.21)$$

Example 4.4

The transmissivity "T" of an aquifer is "approximately 4 m²/min" and the hydraulic gradient "i" is "around 0.02." Calculate the "approximate aquifer discharge," "Q," throughout its whole thickness from a unit width.

By definition, the required aquifer yield is equal to the multiplication of two fuzzy numbers:

$$\text{``}Q\text{''} = \text{``}T\text{''}\,\text{``}i\text{''} \qquad (4.22)$$

First let us attach two triangular MFs for the transmissivity and hydraulic gradient, respectively, as:

$$\mu_T(x) = \begin{cases} 0 & x < 4 \\ x - 4 & 4 < x < 5 \\ -x + 6 & 5 < x < 6 \\ 0 & x > 6 \end{cases} \qquad (4.23)$$

and

$$\mu_i(x) = \begin{cases} 0 & x < 1 \\ x - 1 & 1 < x < 2 \\ -x + 3 & 2 < x < 3 \\ 0 & x > 3 \end{cases} \qquad (4.24)$$

According to the previous calculations, the confidence interval boundaries for fuzzy sets "T" and "i" after the necessary algebraic calculations become "T" = [α+ 4, $-\alpha$ + 6] and "i" = [α + 1, $-\alpha$ + 3]. Finally, the confidence interval boundaries at level α for fuzzy discharge are:

"Q" = [α + 4, $-\alpha$ + 6](.) [α + 1, $-\alpha$ + 3] = [α^2 + 5α + 4, α^2 – 9α + 18]

By equating the left- and right-hand side boundaries to x, the relationships between the actual fuzzy variable x and α are obtained as x = α^2 + 5α + 4 and x = α^2 – 9α + 18 and the solution of these second-order equations in terms of x leads to:

$$\alpha = \frac{\left(-5 + \sqrt{9 + 4x}\right)}{2}$$

and

$$\alpha = \frac{\left(9 - \sqrt{9 + 4x}\right)}{2}$$

Finally, the fuzzy number "Q" MF can be obtained as:

$$\mu_Q(x) = \begin{cases} 0 & x < 4 \\ \dfrac{-5 + \sqrt{9 + 4x}}{2} & 4 < x < 10 \\ \dfrac{9 - \sqrt{9 + 4x}}{2} & 10 < x < 18 \\ 0 & x > 18 \end{cases} \qquad (4.25)$$

The fuzzy numbers are collectively shown in Figure 4.7.

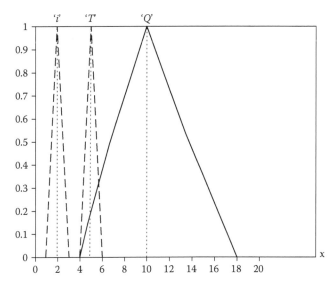

FIGURE 4.7 Fuzzy number multiplication.

It must be noted that the resulting fuzzy number is not triangular because the sides are not straight lines but parabolic (although the curvature is not very clear in this figure).

4.5.1 MULTIPLICATION BY A CONSTANT

Multiplication by a constant changes the scale of the confidence interval by the multiplication of each element between the left and right boundaries. If the confidence interval "X" is multiplied by c, then the resulting confidence interval becomes:

$$\text{"}Z\text{"} = c\cdot\text{"}X\text{"} = [cx_1, cx_2] \tag{4.26}$$

The same procedure is applied for the multiplication of a fuzzy number, say, "X" by a constant c, which leads to a new fuzzy number "Z":

$$\text{"}Z\text{"} = c\cdot\text{"}X\text{"} = [cx_1{}^\alpha, cx_2{}^\alpha] \tag{4.27}$$

Example 4.5

If the rainfall intensity "I" during a single storm is "about" 10 mm/sec and the runoff coefficient c is 0.35, find per unit area surface runoff "r" amount "approximately." Per unit area runoff is a fuzzy number defined by the following multiplication:

$$\text{"}r\text{"} = c\,\text{"}I\text{"} \tag{4.28}$$

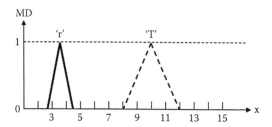

FIGURE 4.8 Fuzzy number multiplication by a constant.

The rainfall intensity fuzzy number can be given as a triangular MF:

$$\mu_I(x) = \begin{cases} 0 & x < 8 \\ \dfrac{x}{2} - 4 & 8 < x < 10 \\ -\dfrac{x}{2} + 6 & 10 < x < 12 \\ 0 & x > 12 \end{cases} \tag{4.29}$$

Because the fuzzy number is multiplied by a constant value, its shape will remain the same as another triangular MF, as in Figure 4.8. The resulting fuzzy number as runoff per unit area is:

$$\mu_r(x) = \begin{cases} 0 & x < 2.8 \\ \dfrac{x}{0.7} - 4 & 2.8 < x < 3.5 \\ -\dfrac{x}{0.7} + 6 & 3.5 < x < 4.2 \\ 0 & x > 4.2 \end{cases} \tag{4.30}$$

4.6 FUZZY DIVISION

The division of two confidence intervals leads to another confidence interval number:

$$\text{``}Z\text{''} = \text{``}X\text{''}/\text{``}Y\text{''} = [x_1/y_2, x_2/y_1] \tag{4.31}$$

This shows that the left (right) boundary of the resulting confidence interval is equal to the division of the left (right) boundary of the first confidence interval to the right (left) boundary of the second confidence interval. The two fuzzy numbers will have the same form of division at α level as:

$$\text{``}Z\text{''} = \text{``}X\text{''}/\text{``}Y\text{''} = [x_1{}^\alpha/y_2{}^\alpha, x_2{}^\alpha/y_1{}^\alpha] \qquad (4.32)$$

Example 4.6

In an area there are "about 100 wells" that withdraw groundwater for water supply to a city with a daily "approximate water demand of 4000 m³." Estimate the "possible discharge" from each well. Approximate well number "n," total demand "D," and the discharge "Q" from each well are fuzzy numbers. The desired discharge value can be calculated as:

$$\text{``}Q\text{''} = \text{``}D\text{''}/\text{``}n\text{''} \qquad (4.33)$$

which means fuzzy number division by another fuzzy number. First, the MFs of given fuzzy numbers are adopted as triangular shapes with the mathematical expressions

$$\mu_D(x) = \begin{cases} 0 & x < 3800 \\ \dfrac{x}{200} - 19 & 3800 < x < 4000 \\ -\dfrac{x}{200} + 21 & 4000 < x < 4200 \\ 0 & x > 4200 \end{cases} \qquad (4.34)$$

and

$$\mu_n(x) = \begin{cases} 0 & x < 80 \\ \dfrac{x}{20} - 4 & 80 < x < 100 \\ -\dfrac{x}{20} + 6 & 100 < x < 120 \\ 0 & x > 120 \end{cases} \qquad (4.35)$$

respectively. For the first fuzzy number, the α-cuts are:

$$\alpha = \left(\dfrac{d_1^\alpha}{200}\right) - 19$$

and

$$\alpha = -\left(\dfrac{d_2^\alpha}{200}\right) + 21$$

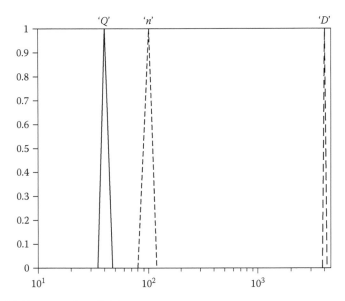

FIGURE 4.9 Fuzzy number division.

Then, "D" α-cut interval values can be calculated as "D" $= [200\alpha + 3800,$ $- 200\alpha + 4200]$. It can be shown in the same manner that "n" $= [20\alpha + 80, - 20\alpha + 120]$. Consequently, the division is:

$$Q_\alpha = D_\alpha /n_\alpha = \left(\frac{200\alpha + 3800}{-20\alpha + 80}, \frac{-200\alpha + 4200}{-20\alpha + 120} \right) \tag{4.36}$$

This can be converted easily into a curve MF function as follows, and the final result is shown in Figure 4.9.

$$\mu_Q(x) = \begin{cases} 0 & x < \dfrac{95}{3} \\[2mm] \dfrac{6x-190}{x+10} & \dfrac{95}{3} < x < 40 \\[2mm] \dfrac{6x-210}{x-10} & 40 < x < \dfrac{105}{3} \\[2mm] 0 & x > \dfrac{105}{3} \end{cases} \tag{4.37}$$

4.6.1 DIVISION BY A CONSTANT

The division of a confidence interval by a constant leads to scale change in the original confidence interval as follows:

$$\text{“}Z\text{”} = \text{“}X\text{”}/c = [x_1/c, x_2/c] \tag{4.38}$$

Likewise, the division of a fuzzy number by a constant c leads to:

$$\text{``}Z\text{''} = \text{``}X\text{''}/c = [x_1^{\alpha}/c, \, x_2^{\alpha}/c] \tag{4.39}$$

Example 4.7

In general, the aquifer transmissivity "T" is calculated through direct field aquifer tests. The aquifer thickness m is found after the subsurface geophysical prospecting "on the average as 10 m." What is the possible hydraulic conductivity "k" value? Suppose that in for a confined aquifer, the transmissivity is "around" 6 m²/hour. First of all, let us represent the transmissivity values as a set of singletons:

$$\text{``}T\text{''} = \left\{ \frac{0.25}{6} + \frac{0.50}{7} + \frac{1}{8} + \frac{0.50}{9} + \frac{0.25}{10} \right\} \tag{4.40}$$

By definition, the transmissivity is equal to the multiplication of hydraulic conductivity by the aquifer thickness:

$$\text{``}T\text{''} = m\text{''}k\text{''} \tag{4.41}$$

or

$$\text{``}k\text{''} = \frac{\text{``}T\text{''}}{m} \tag{4.42}$$

This means that the hydraulic conductivity is a function of transmissivity, that is, "k" = f("T"). Hence, according to Equation (4.41), it is possible to find the hydraulic conductivity fuzzy number by considering the possible transmissivity fuzzy number elements from Equation (2.42).

$$k(6) = \frac{6}{10} = 0.6 \quad k(7) = \frac{7}{10} = 0.7 \quad k(8) = \frac{8}{10} = 0.8 \quad k(9) = \frac{9}{10} = 0.9 \quad k(10) = \frac{10}{10} = 1$$

Hence, the "possible" hydraulic conductivity values become:

$$\text{``}k\text{''} = \left\{ \frac{0.25}{0.6} + \frac{0.50}{0.7} + \frac{1}{0.8} + \frac{0.50}{0.9} + \frac{0.25}{1} \right\}$$

In this example, the dependent variable is a function of one independent variable with MDs.

4.7   EXTREMES OF FUZZY NUMBERS

In many practical applications, it is necessary to calculate extreme (minimum and maximum) values of hydrologic processes. If the data are crisp, then it is easy to identify minimum and maximum in a given interval of time or in an area. For instance, maximum daily rainfall in 2005 is the maximum of 365 daily rainfall amounts in this year.

In the case of fuzzy terminology, the minimum and maximum of two fuzzy numbers is also another fuzzy number, which can be calculated according to fuzzy number minimization ("ANDing") and maximization ("ORing") as explained in Chapter 3. For this purpose, if α-cut level values in "A" and "B" fuzzy numbers are $a_1^{(\alpha)} \leq a_2^{(\alpha)}$ and $b_1^{(\alpha)} \leq b_2^{(\alpha)}$ for all α values (from 0 and 1, inclusive), then the minimization of "A" and "B" sets will be shown as:

$$A_\alpha \wedge B_\alpha = \left\{ \min\left[a_1^{(\alpha)}, b_1^{(\alpha)} \right], \min\left[a_2^{(\alpha)}, b_2^{(\alpha)} \right] \right\} \qquad (4.43)$$

This expression indicates that there are four alternatives in the calculation of α-cuts: $\left[a_1^{(\alpha)}, a_2^{(\alpha)} \right]$, $\left[a_1^{(\alpha)}, b_2^{(\alpha)} \right] \left[b_1^{(\alpha)}, a_2^{(\alpha)} \right]$ and $\left[b_1^{(\alpha)}, b_2^{(\alpha)} \right]$. To find these alternatives, it is necessary to calculate the α values at which the concerned MFs cross each other.

Likewise, maximization operation is given as:

$$A_\alpha \wedge B_\alpha = \left\{ \max\left[a_1^{(\alpha)}, b_1^{(\alpha)} \right], \max\left[a_2^{(\alpha)}, b_2^{(\alpha)} \right] \right\} \qquad (4.44)$$

Example 4.8

Given the following two fuzzy numbers, compute the fuzzy minimum and maximum:

$$\mu_A(x) = \begin{cases} 0 & x \leq -3 \\ x+3 & -3 \leq x \leq -2 \\ \dfrac{-x+5}{7} & -2 \leq x \leq 5 \\ 0 & x > 5 \end{cases}$$

and

$$\mu_B(x) = \begin{cases} 0 & x \leq -4 \\ \dfrac{x+4}{5} & -4 \leq x \leq 1 \\ -x+2 & 1 \leq x \leq 2 \\ 0 & x > 2 \end{cases}$$

As usual, the α-cut of these two fuzzy numbers can be found easily as "A" = $[\alpha - 3, -7\alpha + 5]$ "B" = $[5\alpha - 4, -\alpha + 2]$. According to Equation (4.43), the minimum can be calculated as:

$$A_\alpha \wedge B_\alpha = [(\alpha - 3) \wedge (5\alpha - 4), (-7\alpha + 5) \wedge (-\alpha + 2)]$$

It is now necessary to find α values at which MFs cross each other. For this purpose, the following equalities are considered:

$$\alpha - 3 = 5\alpha - 4$$

and

$$-7\alpha + 5 = -\alpha + 2$$

Simultaneous solution of these last equations leads to the cross point values as $\alpha = 0.25$ and $\alpha = 0.50$. Consideration of these values will be in calculating the desired minimum:

$$A_\alpha \wedge B_\alpha = \begin{cases} [5\alpha - 4, -\alpha + 2] & \text{for} \quad 00.0 \leq \alpha \leq 0.25 \\ [\alpha - 3, -\alpha + 2] & \text{for} \quad 0.25 \leq \alpha \leq 0.50 \\ [\alpha - 3, -\alpha + 2] & \text{for} \quad 0.50 \leq \alpha \leq 1.00 \end{cases}$$

By calculating the appropriate range of the domain variable x, it is possible to calculate the resultant MF from these α-cuts, similar to the procedure used in the fuzzy arithmetic in addition, subtraction, and like operations.

$$\mu_{A_\alpha \wedge B_\alpha} = \begin{cases} 0 & x \leq -4 \\ \dfrac{x+4}{5} & -4 < x \leq -2.75 \\ x+3 & -2.75 < x \leq -2.0 \\ \dfrac{-x+5}{7} & -2.0 < x \leq 1.5 \\ -x+2 & 1.5 < x \leq 2 \\ 0 & x > 2 \end{cases}$$

The minimization MF is shown in Figure 4.10. Likewise, the maximum of the same fuzzy numbers can be calculated from Equation (4.44) along the similar

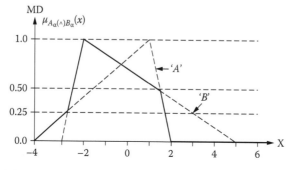

FIGURE 4.10 Minimization MF.

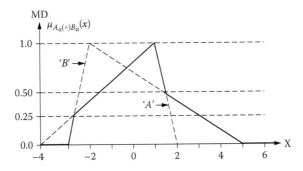

FIGURE 4.11 Maximization MF.

steps given below:

$$A_\alpha \vee B_\alpha = \left[(\alpha-3)\vee(5\alpha-4),(-7\alpha+5)\vee(-\alpha+2)\right]$$

$$A_\alpha \vee B_\alpha = \begin{cases} [\alpha-3,-7\alpha+5] & \text{for} \quad 00.0 \leq \alpha \leq 0.25 \\ [5\alpha-4,-7\alpha+5] & \text{for} \quad 0.25 \leq \alpha \leq 0.50 \\ [5\alpha-4,-\alpha+2] & \text{for} \quad 0.50 \leq \alpha \leq 1.00 \end{cases}$$

Finally, the MF, which is shown in Figure 4.11, becomes:

$$\mu_{A_\alpha(\vee)B_\alpha} = \begin{cases} 0 & x \leq -3 \\ x+3 & -3 < x \leq -2.75 \\ \dfrac{x+4}{5} & 2.75 < x \leq -1.0 \\ -x+2 & 1.0 < x \leq 1.5 \\ \dfrac{-x+5}{7} & 1.5 < x \leq 5 \\ 0 & x > 5 \end{cases}$$

4.8 EXTENSION PRINCIPLE

This is the most useful principle in extending all the mathematical operations into fuzzy subset domain. Given a fuzzy number "A," the purpose is to obtain a fuzzy set "F" through f("A") transformation in discourse domain U. Let "A" be defined explicitly as its singletons.

$$A = \left\{ \frac{\mu_A(x_1)}{x_1} + \frac{\mu_A(x_2)}{x_2} + \cdots + \frac{\mu_A(x_n)}{x_n} \right\}$$

The extension principle states that

$$f(A) = f\left(\frac{\mu_A(x_1)}{x_1} + \frac{\mu_A(x_2)}{x_2} + \cdots + \frac{\mu_A(x_n)}{x_n} \right)$$

$$= \frac{\mu_A(x_1)}{f(x_1)} + \frac{\mu_A(x_2)}{f(x_2)} + \cdots + \frac{\mu_A(x_n)}{f(x_n)}$$

(4.45)

This means that in any transformation, only the values are transformed with the same MDs. The extension principle allows the generalization of crisp mathematical concepts to the fuzzy set framework and extends point-to-point mappings to mapping for fuzzy sets. In this manner, any crisp mathematical relationship between non-fuzzy elements can be extended to deal with fuzzy entities.

Most often, the dependent variable, say "Y," is a function of many independent variables, "X_1," "X_2," ..., "X_n." Let the dependent fuzzy number be "Q" and n independent numbers are $P_1, P_2, \ldots, P_n$ In this case, there will appear more than one MD for the same ordinary number. Then the question is how to choose between these different MDs as the representative for the ordinary value. The answer to this question comes through the extension principle, which states that:

$$\mu_Q(y) = \max \left\{ \min \left[\mu_{P_1}(x_1), \mu_{P_{21}}(x_2), \ldots, \mu_{P_n}(x_n) \right] \right\}$$

$$x_1, x_2, \ldots, x_n$$

(4.46)

$$y = f(x_1, x_2, \ldots, x_n)$$

where Q is a fuzzy set defined over set Y.

Example 4.9

A dam "D" and groundwater resources "G" supply a city. Hence, the total water supply "T" to this city is:

$$"T" = "D" + "G" \quad (4.47)$$

which means that the dependent variable "T" is a function of two independent variables, "D" and "G." If the "approximate" water supplies are given by the following fuzzy singleton numbers,

$$"D" = \left\{ \frac{0.2}{5} + \frac{0.4}{7} + \frac{0.6}{9} + \frac{1.0}{10} + \frac{0.7}{12} + \frac{0.3}{14} \right\}$$

TABLE 4.2

Summations with Extension Principle

"D" \ "G"	1	2	3
5	6	7	8
	(0.3, 0.2)	(1, 0.2)	(0.7, 0.2)
7	8	9	10
	(0.3, 0.4)	(1, 0.4)	(0.7, 0.4)
9	10	11	12
	(0.3, 0.6)	(1, 0.6)	(0.7, 0.6)
10	11	12	13
	(0.3, 1)	(1, 1)	(0.7, 1)
12	13	14	15
	(0.3, 0.7)	(1, 0.7)	(0.7, 0.7)
14	15	16	17
	(0.3, 0.3)	(1, 0.3)	(0.7, 0.3)

and

$$"G" = \left\{ \frac{0.3}{1} + \frac{1.0}{2} + \frac{0.7}{3} \right\}$$

calculate "possible demand" levels.

According to the extension principle, the summation of corresponding "D" and "G" values will be taken but these values have different MDs. The dam provides six different water amounts with "approximate supply of 10 m³" per hour. The groundwater aquifer provides three alternatives "about 2 m³/hour." Hence, the summation of "D" and "G" provides 6 × 3 = 18 different alternatives. The summation is given in Table 4.2.

Herein, the first row and column include the possible values of groundwater and dam supply, respectively. In other cells of the table, the corresponding summation of "D" and "G" alternatives are given with attached MDs in parentheses. The two water supplies can be added, provided they are both available. Because "ANDing" connects them, the selection must be made through a minimization (min) operation (Chapter 3). According to the extension principle in Equation (4.44), the MDs in parentheses are minimized for the representative MDs. The result is presented in Table 4.3.

The total water supply has alternatives of 6, 7, 8, 9, 10, 11, 12, 13, 14, 15, 16, and 17 m³/day, as is obvious from this table. However, some of these values occur more than once with different MDs. For instance, in the table 15 m³/day appears twice with MDs 0.7 and 0.3. Because either 0.3 or 0.7 can be considered, the selection is made by the "ORing" procedure (see Chapter 3). Accordingly, the maximization (max) operation is applied between different MDs. This gives the result of 15 m³/day with 0.7 MD. A similar procedure is applied for the other cases also, and hence the final fuzzy number of total water supply is:

$$"T" = \left\{ \frac{0.2}{6} + \frac{0.2}{7} + \frac{0.3}{8} + \frac{0.4}{9} + \frac{0.4}{10} + \frac{0.6}{11} + \frac{1}{12} + \frac{0.7}{13} + \frac{0.7}{14} + \frac{0.7}{15} + \frac{0.3}{16} + \frac{0.3}{17} \right\}$$

TABLE 4.3
Minimization Procedure of the Extension Principle

"D" \ "G"	1	2	3
5	6	7	8
	(0.2)	(0.2)	(0.2)
7	8	9	10
	(0.3)	(0.4)	(0.4)
9	10	11	12
	(0.3)	(0.6)	(0.6)
10	11	12	13
	(0.3)	(1)	(0.7)
12	13	14	15
	(0.3)	(0.7)	(0.7)
14	15	16	17
	(0.3)	(0.3)	(0.3)

This fuzzy number says that the total water supply to the city is "approximately" 12 m³/day. Within the same fuzzy number the second preference can be adopted as 13 m³/day, 14 m³/day, and 15 m³/day.

REFERENCES

Kaufmann, A. and Gupta, M.M. 1985. *Introduction to Fuzzy Arithmetic*. Van Nostrand Reinhold, New York.

PROBLEMS

4.1 Specify the numbers implied by the following words:
   (a) "several"
   (b) "few"
   (c) "noon"
   (d) "young"
   (e) "porosity"
4.2 Disintegrate "discharge" into at least three categories in descending order from the following words:
   (a) "low"
   (b) "intense"
   (c) "high"
   (d) "small"
   (e) "weak"
   (f) "dry"

4.3 The fuzzy linguistic terms of "high flow," "low flow," and "low rainfall," "high rainfall" are given as follows:

$$\text{``high flow''} = \left\{ \frac{0}{15} + \frac{0.2}{30} + \frac{0.4}{45} + \frac{0.6}{60} + \frac{0.8}{75} + \frac{1.0}{90} \right\}$$

$$\text{``low flow''} = \left\{ \frac{1.0}{15} + \frac{0.8}{30} + \frac{0.6}{45} + \frac{0.4}{60} + \frac{0.2}{75} + \frac{0}{90} \right\}$$

$$\text{``low rainfall''} = \left\{ \frac{1.0}{3} + \frac{0.8}{6} + \frac{0.6}{9} + \frac{0.4}{12} + \frac{0.2}{15} + \frac{0}{18} \right\}$$

$$\text{``high rainfall''} = \left\{ \frac{0}{3} + \frac{0.2}{6} + \frac{0.4}{9} + \frac{0.6}{12} + \frac{0.8}{15} + \frac{1.0}{18} \right\}$$

(a) Find the following membership functions:
   (i)   "very high flow"
   (ii) "not very low rainfall"
   (iii) "not high rainfall" and "very low flow"
   (iv) "slightly low flow"
   (v)   "high rainfall" or "slightly low flow"

4.4 Generally, the rainfall amount (in millimeters per hour) is described linguistically. If the "heavy" and "light" rainfall amounts are given by the following linguistic fuzzy sets, then provide the fuzzy membership functions for the following phrases:

$$\text{``heavy''} = \left\{ \frac{0.2}{4} + \frac{0.4}{7} + \frac{0.6}{11} + \frac{0.8}{19} + \frac{1.0}{29} \right\}$$

$$\text{``light''} = \left\{ \frac{0}{29} + \frac{0.1}{19} + \frac{0.5}{11} + \frac{0.8}{7} + \frac{1.0}{4} \right\}$$

(a) "very heavy"
(b) "slightly heavy"
(c) "fairly heavy," [Note: "fairly heavy" = ("heavy")$^{2/3}$]
(d) "not very light"
(e) "very very fairly light"

4.5 In many hydrologic studies, risk and reliability play a significant role and their domain of variation is from 0 to 1, where 0 corresponds to complete reliability and 1 to risk. Fuzzy risk and reliability statements are given as follows:

$$\text{``risk''} = \left\{ \frac{0}{0} + \frac{0}{0.1} + \frac{0.2}{0.2} + \frac{0.4}{0.3} + \frac{0.6}{0.4} + \frac{0.8}{0.5} + \frac{1.0}{0.6} + \frac{1.0}{0.7} + \frac{1.0}{0.8} + \frac{1.0}{0.9} + \frac{1.0}{1.0} \right\}$$

$$\text{``reliability''} = \left\{ \frac{1.0}{0} + \frac{0.8}{0.1} + \frac{0.5}{0.2} + \frac{0.3}{0.3} + \frac{0.2}{0.4} + \frac{0}{0.5} + \frac{0}{0.6} + \frac{0}{0.7} + \frac{0}{0.8} + \frac{0}{0.9} + \frac{0}{1.0} \right\}$$

Calculate the following linguistic variables:
(a) "very reliable"
(b) "not very reliable"
(c) "fairly risky"
(d) "neither reliable nor risky"

4.6 In a river, there is sedimentation and its scale is described by two words—"large" and "small"—which are fuzzy sets:

$$\text{"large"} = \left\{ \frac{0}{0} + \frac{0.1}{5} + \frac{0.3}{10} + \frac{0.5}{15} + \frac{0.7}{20} + \frac{1.0}{25} \right\}$$

$$\text{"small"} = \left\{ \frac{1.0}{0} + \frac{0.8}{5} + \frac{0.5}{10} + \frac{0.3}{15} + \frac{0.1}{20} + \frac{0}{25} \right\}$$

What are the membership functions of the following linguistic terms?
(a) "very small" and "fairly large"
(b) "not small" or "not large"
(c) "not large" or "small"

4.7 In a groundwater study, the seepage velocity has been described by "fast" and "slow" flow with the following membership function:

$$\text{"fast"} = \left\{ \frac{0}{0} + \frac{0.1}{2} + \frac{0.3}{4} + \frac{0.5}{6} + \frac{0.7}{8} + \frac{1.0}{10} \right\}$$

$$\text{"slow"} = \left\{ \frac{1.0}{0} + \frac{0.9}{2} + \frac{0.7}{4} + \frac{0.5}{6} + \frac{0.2}{8} + \frac{0}{10} \right\}$$

According to these fuzzy statements, answer the following requests:
(a) "not very fast" and "slightly slow"
(b) "very very fast" or "not slightly slow"
(c) "very slow" or "not fast"

4.8 In a catchment, long-term records have indicated that there are two states as "dry" and "wet" according to the following fuzzy sets:

$$\text{"dry"} = \left\{ \frac{1.0}{0} + \frac{0.8}{25} + \frac{0.6}{50} + \frac{0.4}{75} + \frac{0.2}{100} + \frac{0}{125} \right\}$$

$$\text{"wet"} = \left\{ \frac{0}{0} + \frac{0.1}{25} + \frac{0.3}{50} + \frac{0.5}{75} + \frac{0.7}{100} + \frac{1.0}{125} \right\}$$

Find the following linguistic terms:
(a) "slightly wet" and "very dry"
(b) "not very wet" or "rather dry"
(c) "not slightly dry" and "rather wet"

4.9 For sowing wheat, the average temperature requirement is about 8°C. On this basis, "hot" and "cold" linguistic words are expressed numerically as:

$$\text{"hot"} = \left\{ \frac{0}{1} + \frac{0.3}{3} + \frac{0.7}{5} + \frac{0.9}{7} + \frac{1.0}{9} + \frac{0.7}{11} \right\}$$

$$\text{"cold"} = \left\{ \frac{1.0}{1} + \frac{0.8}{3} + \frac{0.6}{5} + \frac{0.4}{7} + \frac{0.2}{9} + \frac{0.1}{11} \right\}$$

Calculate the following membership:
(a) "not very hot"
(b) "slightly cold" or "slightly hot"
(c) "not very cold" and " fairly hot"

4.10 The approximate radius r of a large-diameter well is given according to the following membership function:

$$\text{"}r\text{"} = \left\{ \frac{0.1}{0.5} + \frac{0.4}{0.9} + \frac{0.6}{1.2} + \frac{09}{1.7} + \frac{1.0}{2.1} + \frac{0.1}{3.0} \right\}$$

Find the following linguistic sentences in terms of fuzzy numbers:
(a) The perimeter of the well
(b) The area of the well
(c) If the drawdown during the very early time is about "drawdown" $= \left\{ \frac{0.3}{0.1} + \frac{0.7}{0.4} + \frac{1.0}{0.5} + \frac{0.1}{0.7} \right\}$ m, what is the volume of abstracted water?

4.11 The surface area of a lake is "area" $= \left\{ \frac{0.2}{20} + \frac{0.8}{30} + \frac{1.0}{40} + \frac{0.2}{50} + \frac{0.1}{70} \right\}$ km² and the inflow and outflow into this lake during June are:

$$\text{"inflow"} = \left\{ \frac{0.1}{0.3} + \frac{0.7}{0.4} + \frac{1.0}{0.5} + \frac{0.3}{0.6} + \frac{0.1}{0.7} \right\} \text{ m}^3/\text{sec}$$

and

$$\text{"outflow"} = \left\{ \frac{0.3}{0.1} + \frac{0.5}{0.2} + \frac{1.0}{0.3} + \frac{0.3}{0.4} + \frac{0.1}{0.5} \right\} \text{ m}^3/\text{sec},$$

respectively. Monthly rainfall and evaporation heights are

$$\text{"rainfall"} = \left\{ \frac{0.3}{15} + \frac{1.0}{30} + \frac{0.2}{45} + \frac{0.1}{60} \right\} \text{ mm}$$

and

$$\text{"evaporation"} = \left\{ \frac{0.2}{15} + \frac{1.0}{30} + \frac{0.1}{45} \right\} \text{ mm},$$

respectively. The amount of infiltration during the same month is approximately

$$\text{"infiltration"} = \left\{ \frac{0.2}{15} + \frac{1.0}{30} + \frac{0.1}{45} \right\} \text{ mm}.$$

Calculate the volume variation during this month.

4.12 The peak discharge Q of a catchment is in direct proportion to the runoff coefficient C, catchment area A, and the rainfall intensity I: $(Q = CIA)$. If the approximate linguistic values of the right-hand side variables are

$$\text{``runoff coefficient''} = \left\{ \frac{0.01}{0.1} + \frac{0.7}{0.2} + \frac{0.9}{0.3} + \frac{1.0}{0.4} + \frac{0.1}{0.5} \right\},$$

$$\text{``rainfall intensity''} = \left\{ \frac{0.1}{15} + \frac{0.5}{30} + \frac{1.0}{45} + \frac{0.3}{60} + \frac{0.1}{75} \right\} \text{ mm}$$

and

$$\text{``area''} = \left\{ \frac{0.3}{5} + \frac{0.5}{20} + \frac{1.0}{35} + \frac{0.5}{50} + \frac{0.1}{65} \right\} \text{ km}^2$$

(a) Calculate the approximate peak discharge value.
(b) Find the value of "slight discharge."

4.13 The free water surface area in a dam in July is about "free water surface area" $= \left\{ \frac{0.1}{100} + \frac{0.7}{200} + \frac{0.9}{300} + \frac{1.0}{400} \frac{0.1}{500} \right\} \text{km}^2$. There is volume decrease of about "volume decrease" $= \left\{ \frac{0.2}{50} + \frac{0.4}{150} + \frac{0.8}{250} + \frac{1.0}{350} + \frac{0.5}{400} + \frac{0.1}{500} \right\} x10^6 \text{ m}^3$. The monthly evaporation rate is "evaporation" $= \left\{ \frac{0.2}{15} + \frac{1.0}{30} + \frac{0.1}{45} \right\}$ cm, and the average discharge from the dam is "discharge" $= \left\{ \frac{0}{0} + \frac{0.1}{25} + \frac{0.3}{50} + \frac{0.5}{75} + \frac{0.7}{100} + \frac{1.0}{125} \right\}$ m^3/sec. What is the average discharge at the dam entrance in this month?

4.14 Solar irradiation from the sun is dispersed in approximate percentages directly

$$\text{``direct''} = \left\{ \frac{0}{5\%} + \frac{0.3}{10\%} + \frac{0.7}{15\%} + \frac{1.0}{20\%} + \frac{0.7}{25\%} + \frac{0.1}{30\%} \right\},$$

by clouds

$$\text{``cloud''} = \left\{ \frac{0}{3\%} + \frac{0.2}{8\%} + \frac{0.6}{13\%} + \frac{1.0}{18\%} + \frac{0.5}{22\%} + \frac{0.1}{27\%} \right\},$$

by atmosphere

$$\text{``atmosphere''} = \left\{ \frac{0.1}{2\%} + \frac{0.7}{4\%} + \frac{1.0}{6\%} + \frac{0.6}{8\%} + \frac{0.1}{10\%} \right\}$$

and absorbed by water molecules and air at about

$$\text{``absorption''} = \left\{ \frac{0}{4\%} + \frac{0.5}{8\%} + \frac{0.7}{12\%} + \frac{1.0}{16\%} + \frac{0.3}{18\%} + \frac{0.1}{20\%} \right\}.$$

Accordingly, calculate the average albedo in the northern hemisphere.

4.15 The cross-sectional area at a control section on a river is approximately

$$\text{``area''} = \left\{ \frac{0}{5} + \frac{0.3}{10} + \frac{0.7}{15} + \frac{0.1}{20} + \frac{0.7}{25} + \frac{0.1}{30} \right\} \text{m}^2$$

and the average velocity is given as

$$\text{``velocity''} = \left\{ \frac{0.1}{1} + \frac{0.4}{2} + \frac{0.6}{3} + \frac{0.8}{4} + \frac{1.0}{5} + \frac{0.1}{6} \right\} \text{ m/sec.}$$

Hence, calculate:
(a) "average" discharge
(b) "fairly average" discharge

4.16 If the two fuzzy numbers A and B are given on the real line as follows:

$$\text{``}A\text{''} = \left\{ \frac{0.0}{0} + \frac{0.1}{1} + \frac{0.6}{2} + \frac{0.8}{3} + \frac{0.9}{4} + \frac{0.7}{5} + \frac{0.1}{6} + \frac{0.0}{7} \right\}$$

and

$$\text{``}B\text{''} = \left\{ \frac{0.0}{0} + \frac{1.0}{1} + \frac{0.7}{2} + \frac{0.5}{3} + \frac{0.2}{4} + \frac{0.1}{5} + \frac{0.0}{6} + \frac{0.0}{7} \right\}$$

then calculate the following functional relationships:

(a) $\text{``}C\text{''} = 5(\text{``}A\text{''}) - 3$

(b) $\text{``}C\text{''} = 7.5 \times \sqrt{\text{``}A\text{''}} - \text{``}B\text{''}$

(c) $\text{``}C\text{''} = (\text{``}A\text{''})^2 + (\text{``}B\text{''})^2$

(d) $\text{``}C\text{''} = (\text{``}A\text{''})^{0.7} + (\text{``}A\text{''})(\text{``}B\text{''})$

4.17 Under constant temperature, the fluid's "pressure" and "volume" multiplication is always constant, c. Provided that fuzzy versions of "pressure" and "volume" are given as:

$$\text{``pressure''} = \left\{ \frac{0.0}{0.7} + \frac{1.0}{1.8} + \frac{0.7}{2.2} + \frac{0.5}{3.1} + \frac{0.4}{4.0} \right\}$$

and

$$\text{``volume''} = \left\{ \frac{0.0}{0.5} + \frac{1.0}{0.55} + \frac{0.7}{0.60} + \frac{0.5}{0.65} + \frac{0.0}{0.7} \right\}$$

(a) Using the extension principle, calculate the pressure if the volume is another fuzzy number as:

$$\text{``VOLUME''} = \left\{ \frac{0.0}{0.45} + \frac{0.5}{0.50} + \frac{1.0}{0.55} + \frac{0.5}{0.60} + \frac{0.0}{0.65} \right\}$$

(b) Find "volume"/"pressure" set values.

4.18 Two fuzzy sets A and B are defined by the following membership functions:

$$A(x) = \begin{cases} 1 - \dfrac{|x-5|}{2} & \text{for } 5 \le x \le 7 \\ 0 & \text{for } \text{elsewhere} \end{cases}$$

and

$$B(x) = \begin{cases} 1 - \dfrac{|x-6|}{2} & \text{for} \quad 4 \le x \le 8 \\ 0 & \text{for} \quad \text{elsewhere} \end{cases}$$

(a) Determine the complementary sets of each.
(b) Plot $A(x) \vee B(x)$.

4.19 Let A be a fuzzy set defined on the universal set $\{z, e, k, a, i\}$ as:

$$\text{``}A\text{''} = \left\{ \frac{0.1}{z} + \frac{0.5}{e} + \frac{0.6}{k} + \frac{1.0}{a} + \frac{0.7}{i} \right\}$$

(a) Derive all different α-cuts.
(b) What are the support, hedges, core of "$\overline{A}$".

5 Fuzzy Associations and Clusters

5.1 GENERAL

In the previous chapters, the fuzzy sets and numbers were presented as individual expressions that classify any phenomenon into its subjective and overlapping sub-domains in the forms of MFs. The relationships either between one linguistic variable sub-domain or between two linguistic variables' sub-domains provide key information in all modeling and estimation work. In crisp studies, the correlation coefficient and association indices are used objectively to present the global presence or absence of association, interaction, or interconnectedness between the elements of two crisp sets. These classical methods do not consider the relationships between the elements' sub-domains (fuzzy sets) of the phenomenon, but yield an overall result with some restrictive mathematical assumptions. For instance, the correlation coefficient measures the linear dependence and requires that the data are normally (Gaussian) distributed, stationary, have constant variance (homoscedasticity), etc. (Şen, 1978, 1979, 2001). None of these or any other assumptions are necessary in the fuzzy relationship definitions. Fuzzy relations are generalizations of crisp relations and they allow for various degrees of associations between elements or sub-domains, where degree of association can be represented by MDs and hence the fuzzy relations themselves are fuzzy sets.

To establish a fuzzy relationship, it is necessary that the whole data variation domain be divided into a set of overlapping fuzzy categories with different MFs. Hence, fuzzy sets are related to each other and more refined relationships can be established. Even without data availability experience, expert views and practice may provide a basis for the relationship establishment between different subsets, which are the basis of the fuzzy rules (Chapter 6).

The purpose of this chapter to introduce the basic operations of fuzzy relations through max-min and min-max compositions. It also provides the mutual correspondence and differences between crisp and fuzzy (linguistic) variable relationships and implications in hydrological sciences.

Approximate reasoning is a key motivation in fuzzy sets and possibility theory. This chapter provides a coherent view of this field and its impact on database research and information retrieval. Special emphasis is given to the representation of preliminary fuzzy rules and specialized types of approximate reasoning. Various cluster recognition features through FL approximations are presented.

5.2 CRISP TO FUZZY RELATIONSHIPS

The classical relationship (connectedness) between two sets can be expressed as the presence or absence of association, interaction, or interconnectedness between their elements. Any relationship implies the binary connectedness between two sets, and such relationships cannot be considered for more than two sets at the same time. On the other hand, degree of association can be represented by MDs in a fuzzy binary relation in the same way that the MDs are represented in the fuzzy set. This implies that the binary relations are sets in two dimensions on the Cartesian coordinate diagram in the form of Cartesian product, which can also be shown in matrix form. In crisp sets, this corresponds to the correlation matrix R with correlation elements c_{ij} between sets i and j as follows:

$$R = \begin{bmatrix} 1 & c_{12} & c_{13} & \cdots & c_{1n} \\ c_{21} & 1 & c_{23} & \cdots & c_{2n} \\ c_{31} & c_{32} & 1 & \cdots & c_{3n} \\ \cdot & \cdot & & \cdot\;\cdot\;\cdot & \cdot \\ \cdot & \cdot & & \cdot\;\cdot\;\cdot & \cdot \\ c_{n1} & c_{n2} & c_{n3} & \cdot\;\cdot & 1 \end{bmatrix} \tag{5.1}$$

Herein, by definition, $-1 \leq c_{ij} \leq +1$ and they are crisp values all with characteristic value equal to 1. Calculation of each element, say c_{25}, requires two complete sequences of records or measurements.

Likewise, the association matrix is constructed from the two basic fuzzy sets and their MFs. The fuzzy relationship allows for various degrees of interactions between elements and not between two complete sequences globally. In short, a fuzzy relationship means the grouping of all possible pairs of two sets' elements into convenient fuzzy sets (groups) with certain MFs. To clarify the significance of fuzzy and crisp relationships, it is necessary to glance at Figure 5.1, where a scatter of points is shown from a given set of "X" and "Y" linguistic variables, each categorized into "high" and "low" groups. Let us consider that the horizontal axis indicates historical "streamflow" values versus the previous year's "streamflow" in the same sequence. This is the lag-one scatter diagram, say, of the annual flow series.

The general rule, in surface water hydrology, is that if the correlation coefficient is positive, it implies that "low" flows follow "low" flows and "high" flows follow "high" flows (Figure 5.1a). On the contrary, a negative correlation coefficient is expressed verbally that "low" flows follow "high" flows and "high" flows follow "low" flows (Figure 5.1b). Unfortunately, in many articles and textbooks, these two quotations are stated without considering that they have fuzzy subsets, and they indicate the very basics of fuzzy relationships. The first statement implies a positive

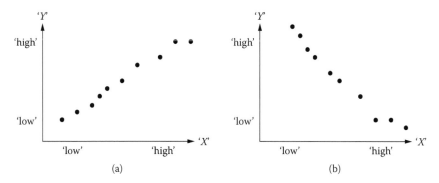

FIGURE 5.1 Lag-one scatter of annual flow series.

correlation in the classical set context but its validity is conditioned on the sequence of "high" ("low") flows with "high" ("low") flows. At this stage, one can notice that the entire streamflow sequence is considered in two categories as "high" and "low." Similar categorizations, preferably with more than two groups, are possible. The classical correlation coefficient definition includes fuzzification ("high" and "low" categories) without refined fuzzy assessment. Such a categorization has the following discrepancies:

1. Because it is based on two categories only, it cannot represent nonlinear relationships, and therefore the correlation coefficient in classical statistical works yields the linear dependence. This is due to the fact that the above quotations do not say anything about the "medium" flow values, which implies the Law of the Excluded Middle of CL (see Chapter 1).
2. The classical quotations with two groups only do not say anything about the categorizations. For instance, in Figure 5.2, although different scatters express the quotation that the "low" ("high") flows follow "low" ("high") flows, close inspection reveals additional information to the reader. The information is not discussed here because each reader will have his or her

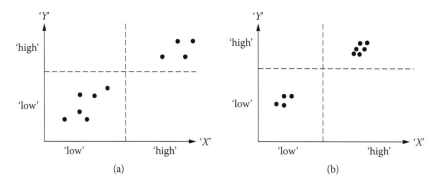

FIGURE 5.2 Annual flow series lag-one scatters (two categories).

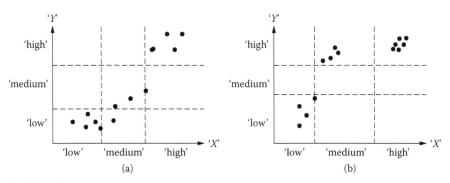

FIGURE 5.3 Annual flow series lag-one scatters with three categories.

own perception and interpretation—although many of them will appear in common.

3. As long as the classical quotation is valid, the correlation coefficient is positive or negative with different values between +1 and 0 and between 0 and −1 corresponding to "direct proportionality" and "indirect proportionality," respectively. Although the numerical correlation coefficient values make a distinction between the strength of correlation, they still exclude the middle (medium) categorization due to CL concepts. However, in the FL consideration, at least three categorizations (Figure 5.3) help make further linguistic nonnumerical relationship statements.

Take a look at Figure 5.4 with different categories and make linguistic relationships between the various variables. Only Figures 5.4a and d are interpreted with relevant relationships using fuzzy concepts in linguistic terms, and the other two are left for the readers' own visualization, interpretation, and deduction. In all the figures, the linear correlation coefficient solutions (regression) are shown as straight lines so that the reader can compare the classical and fuzzy relationship distinctions.

In Figure 5.4a, depending on three categories (fuzzy sets) between the rainfall and runoff, the following three logically valid relationships can be stated linguistically:

1. "low" rainfall implies "low" runoff
2. "medium" rainfall causes mostly "low" runoff
3. "medium" rainfall also gives rise to "high" runoff

As far as the rainfall is concerned, the last two verbal relationships are overlapping fuzzy implications but they have distinctive runoff descriptions. However, the first two relationships have a common runoff category, which is attached to different rainfall categories. These three linguistic statements imply a nonlinear relationship between the rainfall and runoff. Such distinctions are not possible with classical and numerical crisp correlation coefficients.

In Figure 5.4d, again on the basis of three fuzzy subsets for "X" and "Y" variables, the following deductions can be obtained:

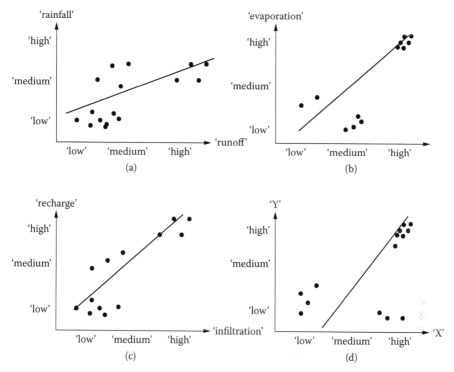

FIGURE 5.4 Hydrologic fuzzy relationships.

1. "low" "X" subset is attached with "low" "Y" category
2. "medium" "X" and "Y" categories are not distinctively related to each other
3. "high" "X" subset implies either "low" or "high" "Y" categories

The reader could get the impression from these statements that there are three distinctive categories between "X" and "Y." Hence, the classical correlation coefficient cannot represent the real situation. The aforementioned explanations hopefully give the reader an appreciation for the relationship between the CL and the linguistic FL statements that relate one variable fuzzy set to another. The reader may wonder that modeling the linguistic statements are not helpful and they must be numerically represented for inclusion in any modeling approach. This is a right path of thinking and in the following sections the numerical representation of these linguistic relationships will be presented through fuzzy concepts.

5.3 LOGICAL RELATIONSHIPS

It is illuminating to look at various hydrological equations from a logical relationship point of view so that the reader may feel more comfortable in transition from classical to fuzzy relationships. In all analytical and empirical equations, CL relationships are hidden. Unfortunately, many hydrologists do not care about the logical relationships but prefer symbolic relationships, that is, mathematical relationships.

In this manner, creative ability and analytical thinking capacity are smoothed (not sharpened) throughout their career with weak innovations. In hydrology, the simple and well-known rational formulation of peak discharge (Q) depends on the CL reasoning, which leads to the following mathematical expression:

$$Q = CIA \qquad (5.2)$$

where A is the catchment area, I is the rainfall intensity, and C is the runoff coefficient. In the modeling sense, Q is the output (effect) variable, whereas C, I, and A are input (cause) variables. Similar to other formulations in the hydrology literature, this is a CL conclusion that can be decomposed into the following logical statements:

1. As the catchment area A increases, the peak discharge Q increases.
2. Likewise, as the rainfall intensity I increases, the peak discharge increases.
3. As the runoff coefficient C increases, the peak discharge estimate also increases.
4. As any of the two variables (C, I, A) increase, the peak discharge increases.
5. As three of the input variables increase, the peak discharge increases.

It is interesting to note that in all these logical relationships, there is a positive correlation (direct proportionality) similar to the quotation in the previous section concerning the streamflow records, and therefore none of the variables correspond to classically linear (directly or indirectly proportional) implications.

To interpret the same formulation by an FL approach and relationships, it is first necessary to consider each variable in the form of at least three overlapping fuzzy sets. For instance, each linguistic variable in Equation (5.2) can be thought to have the following adjectives (see Chapter 2):

1. Discharge: "low," "medium," "high," and "extremely high"
2. Area: "small," "medium," and "big"
3. Rainfall intensity: "low," "moderate," and "high"
4. Runoff coefficient: "small," "medium," and "big"

Consideration of these sub-categories increases the number of relationship statements between the discharge and the other variables. For instance, instead of a single statement of a global crisp relationship such that as the area increases, the peak discharge increases, there will be $4 \times 3 = 12$ inter-subset relationship statements in FL terminology as follows:

1. What is the relationship between the "low" discharge and "small" catchment area?
2. What is the relationship between the "low" discharge and "medium" catchment area?

3. What is the relationship between the "low" discharge and "big" catchment area?
4. What is the relationship between the "medium" discharge and "small" catchment area?
5. What is the relationship between the "medium" discharge and "medium" catchment area?
6. What is the relationship between the "medium" discharge and "big" catchment area?
7. What is the relationship between the "high" discharge and "small" catchment area?
8. What is the relationship between the "high" discharge and "medium" catchment area?
9. What is the relationship between the "high" discharge and "big" catchment area?
10. What is the relationship between the "extremely high" discharge and "small" catchment area?
11. What is the relationship between the "extremely high" discharge and "medium" catchment area?
12. What is the relationship between the "extremely high" discharge and "big" catchment area?

These are the bivariate fuzzy relationship statements. However, it is possible to write relationship by triple fuzzy sets, say catchment area and rainfall intensity effect on the discharge. Even quadruple relationship statements can be written, if necessary, but this is left to readers for their practical training.

The hydrologist has the opportunity to think in more detail in order to answer each of the above questions. Hence, it is obvious that FL relationship thinking provides detailed information and a more refined approach; this needs expert views and then numerical allocations such as MDs through fuzzy sets for fixation of the answers.

5.4 FUZZY LOGIC RELATIONS

In the case of two crisp sets U and V, the Cartesian product of U and V, which is denoted by $U \times V$, is the set of all ordered member pairs (u, v) such that:

$$U \times V = \{(u, v) \mid u \in U \text{ and } v \in V\} \tag{5.3}$$

In this product, the order is important; therefore, $U \times V \neq V \times U$. This indicates all the possible members in the relationship of the whole two universal sets. The subset relationship is also included in the final universal product as another subset. If the relation is shown by $R(U, V)$, then

$$R(U, V) \subset U \times V \tag{5.4}$$

where $R(U, V)$ can be expressed linguistically, such as "members greater than 5," "members between 4 and 11," "members smaller than 10," etc. Each one of these linguistic expressions implies a certain relationship.

Example 5.1

Given $U = \{2, 5, 9\}$ and $V = \{0, 3, 6\}$ crisp sets, the Cartesian product results in $U \times V = \{(2, 0), (2, 3), (2, 6), (5, 0), (5, 3), (5, 6), (9, 0), (9, 3), (9, 6)\}$. If the relationship is expressed linguistically as "the first member is smaller than the second member," then the given universal set will have members that express this relationship and with their collection the relationship subset is obtained as:

$$R(U, V) = \{(2,3), (2, 6), (5, 6)\}$$

Because the resultant relation is in the form of a set, all the set operations can be applied without restriction. In the case of crisp sets, the CV of the relationship set $\mu_R(.)$ can be written as:

$$\mu_R(u_1, u_2, \ldots, u_n) = \begin{cases} 1 & \text{if} \quad (u_1, u_2, \ldots, u_n) \in R(U_1, U_2, \ldots, U_n) \\ 0 & \text{otherwise} \end{cases} \tag{5.5}$$

It is also possible to show the relationships in the form of matrix, which is then referred to as the relation matrix similar to the matrix in Equation (5.1). For the above example, the relation matrix can be written as in Table 5.1 with corresponding matrix form.

For classical sets, the relationship degree will be crisp as either 1 or 0. For certain relationships, it is difficult to give a zero-one assessment where the fuzzy relations come into view.

Example 5.2

Let the two universal sets of different drainage basins be that from the Middle East as $U = \{Euphrates, Tigris, Nile\}$ and from Europe as $V = \{Danube, Rhine\}$. It is possible to relate these two sets of drainage basins according to distances, say through the fuzzy words *very far*. This linguistic word cannot be treated as a crisp set, and therefore its MDs will vary between 0 and 1, inclusively. Consideration of

TABLE 5.1
Crisp Relation Matrix

		V		
		0	3	6
U	2	0	1	1
	5	0	0	1
	9	0	0	0

$$R(U, V) = \begin{bmatrix} 0 & 1 & 1 \\ 0 & 0 & 1 \\ 0 & 0 & 0 \end{bmatrix}$$

TABLE 5.2
'very far' Relation Matrix

		V	
		Danube	Rhine
	Euphrates	0.7	0.9
U	Tigris	0.8	1.0
	Nile	0.5	0.6

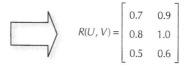

$$R(U, V) = \begin{bmatrix} 0.7 & 0.9 \\ 0.8 & 1.0 \\ 0.5 & 0.6 \end{bmatrix}$$

these five drainage basins will give approximate answers to the distances between them. Hence, the concept of "very far" can be represented numerically in a matrix relationship as in Table 5.2 with its matrix form. This example shows that it is necessary to generalize the concept of classical relationship in order to formulate more relationships in the real world.

Example 5.3

If U and V represent the set of real numbers, then a fuzzy relationship as "x is approximately equal to y" may be defined by the following MF:

$$\mu_{UV}(x, y) = \frac{1}{e^{-(x-y)^2}} \tag{5.6}$$

Likewise, if a fuzzy relationship says that "x is much larger than y," then its MF can be represented as:

$$\mu_{UV}(x, y) = \frac{1}{1 + e^{-(x-y)}} \tag{5.7}$$

These two examples indicate that, at times, there is a mathematical representation possibility for the fuzzy relationships. Hence, in our daily conversations, there may be many statements that imply linguistic relationships that can be translated into mathematical formulations.

In the classical concept of relationship, there is a crisp decision that the objects are either dependent or independent. Furthermore, the dependence is measured by a crisp number (between −1 and +1), such as the classical correlation coefficient, which shows the strength of the dependence between two sets as in Equation (5.1). Whatever the correlation strength, the final decision is in the form of a relationship. The generalization of the crisp relationship allows for various degrees of interaction between elements, and these degrees imply fuzzy relations. The basic philosophy behind a fuzzy relation is that everything is either related completely or partially or unrelated. There may be confusion at this stage concerning the strength of relationship for crisp sets and MD for fuzzy sets. To make the first distinction between these two types of relations, it is useful to think about the crisp relationship numerically, and the fuzzy relations in terms of words (fuzzy sets), linguistically. Although a crisp relationship is expressed with a number, say 0.8, which is the strength that shows its nearness to a complete relation represented by 1, this number does not say anything about the relationship between classified sub-groups of two sets. For instance, the crisp relation number cannot answer the questions such as:

1. What is the relation between a specific element of set U and the corresponding element in set V? That is, the classical relation cannot express the relation on an element basis.
2. What is the linguistic relation between specified subsets of set U—for example, "low discharge" with "low rainfall"?

On the other hand, given the two crisp sets—$U = \{1, 2.5, 4.3, 0.8, 3.2\}$ and $V = \{1.98, 5.07, 9.54, 1.61, 6.26\}$—their relations can be identified after the plot of set U versus set V elements as shown in Figure 5.5. It is obvious that although there is a scatter of data points on the Cartesian coordinate system, there is an increasing trend, which implies a directly proportional (positive) relationship. It can be expressed linguistically as "low" ("high") U element values follow "low" ("high") V values with the Law of the Excluded Middle principle.

On the contrary, the scatter points in Figure 5.6 do not reveal any relation between two given sets (A and B) because the points are scattered in a random manner all over the area without any trend observation. In these scatter diagrams, all the points have CV equal to 1 and non-points have CV equal to zero.

For a fuzzy relationship, the two fuzzy sets "U" and "V" are given with the MDs as:

$$"U" = \left\{ \frac{1.0}{1} + \frac{0.9}{2.5} + \frac{0.5}{4.3} + \frac{0.1}{0.8} + \frac{0}{3.2} \right\}$$

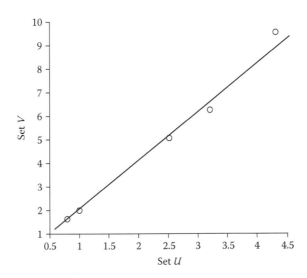

FIGURE 5.5 Crisp relationship.

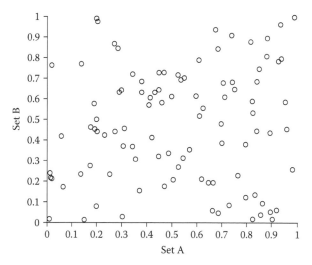

FIGURE 5.6 Crisp relation nonexistence.

and

$$"V" = \left\{ \frac{0.5}{1.98} + \frac{0.7}{5.07} + \frac{0.95}{9.54} + \frac{1.0}{1.61} + \frac{1.0}{6.26} \right\}$$

By considering only the support values (without MDs) and their simultaneous occurrences, the scatter diagram appears as dots in circles only (Figure 5.7). However, if all the alternative pairs are considered, the scatter diagram is as dots in Figure 5.7. Consideration of MDs with these dots gives rise to a question concerning each scatter point in Figure 5.7 with two different MDs as to which one to take as the representation of the point? It is logical that the smaller MD

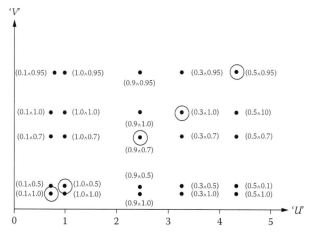

FIGURE 5.7 FAM scatter diagram.

must be adopted because any point is representable with two MDs as "x AND y" through the "ANDing" conjunction, which corresponds to the minimization operation of the fuzzy sets (Chapter 3). It is very convenient to represent the fuzzy relationship in the form of a matrix, which is referred to as the fuzzy associative matrix (or memory) (FAM). For the given fuzzy sets "U" and "V," one can write:

$$
R(\text{"}U\text{"}, \text{"}V\text{"}) =
\begin{bmatrix}
1.0 \wedge 0.5 & 1.0 \wedge 0.7 & 1.0 \wedge 0.95 & 1.0 \wedge 1.0 & 1.0 \wedge 1.0 \\
0.9 \wedge 0.5 & 0.9 \wedge 0.7 & 0.9 \wedge 0.95 & 0.9 \wedge 1.0 & 0.9 \wedge 1.0 \\
0.5 \wedge 0.5 & 0.5 \wedge 0.7 & 0.5 \wedge 0.95 & 0.5 \wedge 1.0 & 0.5 \wedge 1.0 \\
0.1 \wedge 0.5 & 0.1 \wedge 0.7 & 0.1 \wedge 0.95 & 0.1 \wedge 1.0 & 0.1 \wedge 1.0 \\
0.0 \wedge 0.5 & 0.0 \wedge 0.7 & 0.0 \wedge 0.95 & 0.0 \wedge 1.0 & 0.0 \wedge 1.0
\end{bmatrix}
$$

where $R(\text{"}U\text{"}, \text{"}V\text{"})$ indicates the relation between the fuzzy sets. In this matrix, each column corresponds to each element in the fuzzy set "U" and each row corresponds to each element in fuzzy set "V." After the necessary operations, this fuzzy relation matrix can be written succinctly as,

$$
R(\text{"}U\text{"}, \text{"}V\text{"}) =
\begin{bmatrix}
0.5 & 0.7 & 0.95 & 1.0 & 1.0 \\
0.5 & 0.7 & 0.9 & 0.9 & 0.9 \\
0.5 & 0.5 & 0.5 & 0.5 & 0.5 \\
0.1 & 0.1 & 0.1 & 0.1 & 0.1 \\
0.0 & 0.0 & 0.0 & 0.0 & 0.0
\end{bmatrix}
$$

This FAM matrix has different properties when compared with the classical correlation matrix in Equation (5.1), including:

1. The FAM indicates interconnectedness between element pairs, whereas Equation (5.1) shows correlation coefficients between two complete sets.
2. The FAM is not a square matrix, which implies that the two sets should not be necessarily of the same size. They can be of different sizes, such as n and m, and correspondingly FAM is an $n \times m$ matrix.
3. The FAM has elements, which are MDs, and they are confined in $0 < MD < 1$ interval rather than $-1 < c_{ij} < +1$ as in Equation (5.1).
4. As a consequence of the previous point, the FAM is not necessarily a symmetrical matrix.
5. There are no special main diagonal elements, whereas in the correlation matrix the main diagonal elements are equal to +1.
6. The FAM values can be shown on a Cartesian coordinate system systematically as in Figure 5.7, which provides the fuzzy relationship scatter on the Cartesian domain of two fuzzy sets.

Any FAM can be considered a two-dimensional fuzzy set (or number) and it is possible to apply α-cut operation like for the fuzzy numbers in Chapter 4. Such an operation reveals a set of crisp relationship matrix between the two sets. For instance, crisp relationship matrices at 0.5, 0.7, and 0.9 α-cut levels, respectively, are:

$$R_{0.5}("U","V") = \begin{bmatrix} 0.5 & 0.7 & 0.95 & 1.0 & 1.0 \\ 0.5 & 0.7 & 0.9 & 0.9 & 0.9 \\ 0.5 & 0.5 & 0.5 & 0.5 & 0.5 \\ 0.0 & 0.0 & 0.0 & 0.0 & 0.0 \\ 0.0 & 0.0 & 0.0 & 0.0 & 0.0 \end{bmatrix}$$

$$R_{0.7}("U","V") = \begin{bmatrix} 0.0 & 0.7 & 0.95 & 1.0 & 1.0 \\ 0.0 & 0.7 & 0.9 & 0.9 & 0.9 \\ 0.0 & 0.0 & 0.0 & 0.0 & 0.0 \\ 0.0 & 0.0 & 0.0 & 0.0 & 0.0 \\ 0.0 & 0.0 & 0.0 & 0.0 & 0.0 \end{bmatrix}$$

and

$$R_{0.7}("U","V") = \begin{bmatrix} 0.0 & 0.0 & 0.95 & 1.0 & 1.0 \\ 0.0 & 0.0 & 0.9 & 0.9 & 0.9 \\ 0.0 & 0.0 & 0.0 & 0.0 & 0.0 \\ 0.0 & 0.0 & 0.0 & 0.0 & 0.0 \\ 0.0 & 0.0 & 0.0 & 0.0 & 0.0 \end{bmatrix}$$

Example 5.4

Consider the FAM between two fuzzy sets "X" and "Y" as:

$$R("X","Y") = \begin{bmatrix} 0.4 & 0.5 & 0.0 \\ 0.9 & 0.5 & 0.0 \\ 0.0 & 0.0 & 0.3 \\ 0.3 & 0.9 & 0.9 \end{bmatrix}$$

It is possible to decompose this matrix into all α-cuts (0.3, 0.4, 0.5, and 0.9) and hence write it as a summation:

$$R("X","Y") = 0.3R_{0.3} + 0.4R_{0.4} + 0.5R_{0.5} + 0.9R_{0.9}$$

$$= 0.3 \begin{bmatrix} 1.0 & 1.0 & 0.0 \\ 1.0 & 1.0 & 0.0 \\ 0.0 & 0.0 & 1.0 \\ 1.0 & 1.0 & 1.0 \end{bmatrix} + 0.4 \begin{bmatrix} 1.0 & 1.0 & 0.0 \\ 1.0 & 1.0 & 0.0 \\ 0.0 & 0.0 & 0.0 \\ 1.0 & 1.0 & 1.0 \end{bmatrix}$$

$$+ 0.5 \begin{bmatrix} 0.0 & 1.0 & 0.0 \\ 1.0 & 1.0 & 0.0 \\ 0.0 & 0.0 & 1.0 \\ 0.0 & 1.0 & 1.0 \end{bmatrix} + 0.9 \begin{bmatrix} 0.0 & 0.0 & 0.0 \\ 1.0 & 0.0 & 0.0 \\ 0.0 & 0.0 & 0.0 \\ 0.0 & 1.0 & 0.0 \end{bmatrix}$$

Example 5.5

There are five small agricultural plots on both sides of a river as shown in Figure 5.8. Right and left banks have area sets as $R = \{r_1, r_2, r_3\}$ and $L = \{l_1, l_2\}$, respectively. These plots are irrigated from the river flow. If the distances between plots on both sides are considered, the first impression from Figure 5.8 may be that the distances need "short" travel time for water to reach each plot from the river. In this sentence, *short* is a fuzzy word, and therefore the relative shortness of distances may be attached to numbers between 0 and 1 as in Figure 5.9. It shows a fuzzy binary relation (sagittal graph) referring to "short" time fuzziness.

Graphs similar to Figure 5.8, which connect the mutual elements of two sets, are referred to as sagittal diagrams. Fuzzy relationships can be shown either in the form of matrices or sagittal diagrams, which show explicitly the connections and the strength of each pair. One can also arrange these distance strengths in matrix form as the FAM:

$$R("r","l") = \begin{array}{c} r_1 \\ r_2 \\ r_3 \end{array} \begin{array}{cc} l_1 & l_2 \\ \begin{bmatrix} 0.3 & 0.9 \\ 0.2 & 0.7 \\ 0.1 & 0.8 \end{bmatrix} \end{array} \qquad (5.8)$$

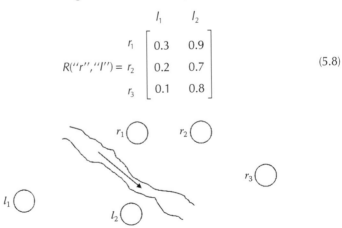

FIGURE 5.8 Agricultural plots distribution.

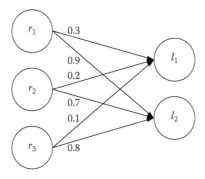

FIGURE 5.9 Fuzzy distance relation.

In general, if there are *n* fuzzy numbers "F_1", "F_2",..., "F_n" defined on sets X_1, X_2,..., X_n, the Cartesian product of these fuzzy numbers is defined symbolically as "F_1" × "F_2" × ... × "F_n." The MF of the Cartesian product of the fuzzy sets is calculated through the minimization operation because the Cartesian coordinates imply "ANDing" (see Chapter 3):

$$\mu_{F_1 \times F_2 \times ..... \times F_n}(x_1, x_2,....., x_n) = \mu_{F_1}(x_1) \wedge \mu_{F_2}(x_2) \wedge ...... \wedge \mu_{F_n}(x_n) \tag{5.9}$$

Example 5.6

There are two rivers R_1 and R_2 in a region with "big discharge" (m³/sec) values given as fuzzy numbers:

$$"R_1" = \left\{ \frac{0.3}{200} + \frac{0.6}{400} + \frac{1}{800} \right\}$$

and

$$"R_2" = \left\{ \frac{0.2}{150} + \frac{0.9}{600} \right\}$$

The problem is to find pairs of rivers with "big" discharges. The solution is the Cartesian product of these two fuzzy sets:

$$"R_1" \times "R_2" = \left\{ \frac{\min(0.3, 0.2)}{(200, 150)} + \frac{\min(0.3, 0.9)}{(200, 600)} + \frac{\min(0.6, 0.2)}{(400, 150)} \right.$$

$$\left. + \frac{\min(0.6, 0.9)}{(400, 150)} + \frac{\min(1, 0.2)}{(800, 150)} + \frac{\min(1, 0.9)}{(800, 600)} \right\}$$

Finally,

$$"R_1" \times "R_2" = \left\{ \frac{0.2}{(200, 150)} + \frac{0.3}{(200, 600)} + \frac{0.2}{(400, 150)} + \frac{0.6}{(400, 150)} + \frac{0.2}{(800, 150)} + \frac{0.9}{(800, 600)} \right\}$$

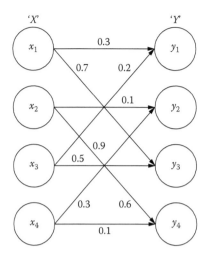

FIGURE 5.10 Simple sagittal graph.

Example 5.7

Consider two fuzzy sets with the following FAM on "X" × "Y" space

$$R("X","Y") = \begin{bmatrix} 0.3 & 0.0 & 0.7 & 0.0 \\ 0.0 & 1.0 & 0.9 & 0.6 \\ 0.2 & 0.0 & 0.5 & 0.0 \\ 0.0 & 0.3 & 0.0 & 1.0 \end{bmatrix}$$

Then the sagittal graph appears as in Figure 5.10.

5.5 FUZZY COMPOSITIONS

Two fuzzy relationships may have a common fuzzy set and hence it is necessary to look for a relationship between two relation matrices. If two relations are R_1("X", "Y") and R_2("Y", "Z"), then the two relations have set "Y" as a common universe of discourses. The composition principle of R_1 and R_2, denoted by $R_1 \circ R_2$, is defined as the relation in "X" × "Z" such that $(x, z) \in R_1 \circ R_2$ if and only if there exists at least one $y \in$ "Y" such that $(x, y) \in$ "R_1" and $(y, z) \in$ "R_2". The composition can be written in terms of MFs as:

$$\mu_{R_1 \circ R_2}(x, z) = \max_{y \in Y} t \left[\mu_{R_1}(x, y), \mu_{R_2}(y, z) \right]$$

for any $(x, z) \in$ "X" × "Z", where t is any t-norm (see Chapter 3). Basically, there are two types of composition operations: maximization-minimization (max-min) and minimization-maximization (min-max) computations. The max-min composition of R_1 and R_2 is defined as:

$$\mu_{R_1 \circ R_2}(x, z) = \max_{y \in Y} - \min t \left[\mu_{R_1}(x, y), \mu_{R_2}(y, z) \right] \tag{5.10}$$

where $(x, z) \in$ "X" $\times$ "Z". The max-min composition of fuzzy relations is given by the following MF:

$$\mu_{R_1 \circ R_2}(x, z) = \min_{y \in Y} - \max t \left[\mu_{R_1}(x, y), \mu_{R_2}(y, z) \right] \tag{5.11}$$

where $(x, z) \in$ "X" $\times$ "Z".

In practical studies, the max-min composition is the most commonly used and indicates that fuzzy relations can be interpreted as indicating the strength of the existence of a relational chain between the elements of "X" and "Z." The max-min composition can be generalized to other compositions by replacing the min operator with any t-norm operator. For instance, max-product (max-prod) composition is given as:

$$\mu_{R_1 \circ R_2}(x, z) = \max_{y \in Y} - prod \, t \left[\mu_{R_1}(x, y) . \mu_{R_2}(y, z) \right] \tag{5.12}$$

Example 5.8

There are three regions, X, Y, and Z with 3, 2, and 4 dams in each, respectively. Hence, the sets of dams can be denoted as $X = \{x_1, x_2, x_3\}$, $Y = \{y_1, y_2\}$, and $Z = \{z_1, z_2, z_3, z_4\}$. The flow is first from X to Y and then from Y to Z. It is therefore possible to make assessments of discharges from dams in region X to Y and Y to Z, but not directly from X to Z. This implies a composition problem. Expert views have provided the "low" discharge possibilities between two successive dams according to the following FAMs:

$$R_1 = \begin{array}{c} \\ x_1 \\ x_2 \\ x_3 \end{array} \begin{array}{cc} y_1 & y_2 \\ \left[\begin{array}{cc} 0.3 & 0.9 \\ 0.2 & 0.7 \\ 0.1 & 0.8 \end{array} \right] \end{array}$$

and

$$R_2 = \begin{array}{c} \\ y_1 \\ y_2 \end{array} \begin{array}{cccc} z_1 & z_2 & z_3 & z_4 \\ \left[\begin{array}{cccc} 1.2 & 0.4 & 0.3 & 0.2 \\ 1.1 & 0.9 & 1 & 0.5 \end{array} \right] \end{array}$$

Fuzzy relation $R_1 \circ R_2$ refers to "low" discharge transfer between the regions of set "X" and "Z" through dams in region Y. The MD calculations for dams x_1 and z_4 in fuzzy relation $R_1 \circ R_2$ are achieved through the minimization of the fuzzy MDs as follows:

$$\min\{\mu_{R_1}(x_1, y_1), \mu_{R_2}(y_1, z_4)\} = \min\{0.3, 0.2\} = 0.2$$

$$\min\{\mu_{R_1}(x_1, y_2), \mu_{R_2}(y_2, z_4)\} = \min\{0.9, 0.5\} = 0.5$$

The composition of these two results can be achieved by the maximization operation as:

$$\mu_{R_1 \circ R_2}(x_1, z_4) = \max(0.2, 0.3) = 0.3$$

The overall composition from two matrices can be obtained similar to matrix multiplication. Two matrices must be compatible; that is, the number of rows in the first matrix must be equal to the number of columns in the second matrix. Corresponding elements between any rows from the first matrix with the elements in any column of the second matrix must be minimized, and hence pairs of corresponding elements are reduced to a sequence of single elements. Finally, the maximization operation on the single sequence of elements gives the final value in the composition matrix. Consideration of the previous two matrices leads to:

$$R_1 \circ R_2 = \begin{bmatrix} 0.3 & 0.9 \\ 0.2 & 0.7 \\ 0.1 & 0.8 \end{bmatrix} \circ$$

$$\begin{bmatrix} 0.2 & 0.4 & 0.3 & 0.2 \\ 0.1 & 0.9 & 1 & 0.5 \end{bmatrix} = \begin{bmatrix} 0.2 & 0.9 & 0.9 & 0.5 \\ 0.2 & 0.7 & 0.7 & 0.5 \\ 0.1 & 0.8 & 0.8 & 0.5 \end{bmatrix} \qquad (5.12)$$

Example 5.9

Reconsider the drainage basin sets from the Middle East and Europe as in Example 5.2. This time, look for another fuzzy relation as the linguistic expression of "very near" drainage basins between the European and Middle East drainage basins. Now the following explicit relationship (FAM) can be written with the members and MDs as:

"P" = 0.7/(Euphrates, Danube) + 0.9/(Euphrates, Rhine) + 0.8/(Tigris, Danube) + 1.0/(Tigris, Rhine + 0.5/(Nile, Danube) + 0.6/(Nile, Rhine)

The new relation from the last matrix is,

"R" = 0.95/(Danube, Indus) + 0.7/(Danube, Yankee) + 0.5/(Rhine, Indus) + 0.3/(Rhine, Yankee)

It is now possible to apply first the max-min, and then max-product operations in order to find the relation between "*P*" and "*R*." "*P*" × "*R*" contains six elements, which are (Euphrates, Indus), (Euphrates, Yankee), (Tigris, Indus), (Tigris, Yankee), (Nile, Indus), and (Nile, Yankee). The main task is to determine the MDs of these six elements. The max-min operator of composition gives:

$$\mu_{P \circ R}(\text{Euphrates, Indus}) = \max\{\min[\mu_P(\text{Euphrates,Danube}), \mu_R(\text{Danube,Indus})],$$
$$\min[\mu_P(\text{Euphrates,Rhine}), \mu_R(\text{Rhine,Indus})]\}$$
$$= \max[\min(0.7, 0.95), \min(0.9, 0.5)]$$
$$= 0.7$$

Similarly,

$$\mu_{P \circ R}(\text{Euphrates,Yankee}) = \max\{\min[\mu_P(\text{Euphrates,Danube}), \mu_R(\text{Danube, Yankee})],$$
$$\min[\mu_P(\text{Euphrates,Rhine}), \mu_R(\text{Rhine, Yankee})]\}$$
$$= \max[\min(0.7, 0.70), \min(0.9, 0.3)]$$
$$= 0.7$$

The reader can continue in a similar fashion to find all other relevant terms. If the max-product composition operator is used, then the following first two fuzzy set members are:

$$\mu_{P \circ R}(\text{Euphrates, Indus}) = \max\{[\mu_P(\text{Euphrates, Rhine})\mu_R(\text{Rhine, Indus})],$$
$$[\mu_P(\text{Euphrates, Danube})\mu_R(\text{Danube, Indus})]\}$$
$$= \max[0.7 \times 0.95, 0.9 \times 0.5]$$
$$= 0.665$$

Similarly,

$$\mu_{P \circ R}(\text{Euphrates,Yankee}) = \max\{[\mu_P(\text{Euphrates,Rhine})\mu_R(\text{Rhine, Yankee})],$$
$$[\mu_P(\text{Euphrates, Danube})\mu_R(\text{Danube, Yankee})]\}$$
$$= \max[0.7 \times 0.7, 0.9 \times 0.3]$$
$$= 0.49$$

Other terms can be computed in the same way by the reader. Another way of accomplishing the compositional calculations is to multiply the relation matrices with consideration of either the max-min or max-product rules. The relational matrix for the fuzzy composition $P \circ R$ can be computed according to the following methods:

1. The max-min operator requires writing out each member in the matrix product *PR* but treating each multiplication as a min operation and each addition as a max operation, and
2. For max-product composition, one should write out each member in the matrix product *PR* but treat each addition as a max operation.

Example 5.10

Perform the previous composition operations through matrix operations. The composition according to the max-min operator can be written explicitly as:

$$
\begin{pmatrix} 0.7 & 0.9 \\ 0.8 & 1.0 \\ 0.5 & 0.6 \end{pmatrix} \circ \begin{pmatrix} 0.95 & 0.70 \\ 0.50 & 0.30 \end{pmatrix} = \begin{pmatrix} 0.7 & 0.7 \\ 0.8 & 0.7 \\ 0.5 & 0.5 \end{pmatrix}
$$

The max-product composition leads to:

$$
\begin{pmatrix} 0.7 & 0.9 \\ 0.8 & 1.0 \\ 0.5 & 0.6 \end{pmatrix} \circ \begin{pmatrix} 0.95 & 0.70 \\ 0.50 & 0.30 \end{pmatrix} = \begin{pmatrix} 0.665 & 0.49 \\ 0.76 & 0.42 \\ 0.475 & 0.35 \end{pmatrix}
$$

Example 5.11

Denote four water resources availability as $W = [W_1, W_2, W_3, W_4]$ for the water supply to an urban area and three experts $E = [E_1, E_2, E_3]$ are consulted for a final decision with different preferences for each water resource. How can the composition of fuzzy relations help the experts in their decision making?

The experts' interests are represented by a fuzzy relation $I("X", "Y")$ given as:

$$
I("X","Y") = \begin{bmatrix} 0.2 & 1.0 & 0.8 & 0.1 \\ 1.0 & 0.1 & 0.0 & 0.5 \\ 0.5 & 0.9 & 0.5 & 1.0 \end{bmatrix}
$$

The properties of water resources are indicated by the fuzzy relation $P("Y", "Z")$ as:

$$
P("Y","Y") = \begin{bmatrix} 1.0 & 0.5 & 0.6 & 0.1 \\ 0.2 & 1.0 & 0.8 & 0.8 \\ 0.0 & 0.3 & 0.7 & 0.0 \\ 0.1 & 0.5 & 0.8 & 1.0 \end{bmatrix}
$$

In light of this information, the max-min composition of $I("X", "Y")$ and $P("Y", "Z")$ can help the experts choose the proper resources according to the following FAM:

$$
I \circ P = \begin{bmatrix} 0.2 & 1.0 & 0.8 & 0.8 \\ 1.0 & 0.5 & 0.6 & 0.5 \\ 0.5 & 0.9 & 0.8 & 1.0 \end{bmatrix}
$$

5.6 LOGICAL CATEGORIZATION

Modeling seeks a meaningful relationship between the input and output parameters of an event for predicting or estimating the output performance of the system. In hydrological modeling, the relationships between pairs of variables play a significant role and, if the relationship type (linear or nonlinear) and the mutual effects between the variables are known through correlation coefficients, then it is possible to construct crisp linear models. However, fuzzy and nonlinear models require measures of a nonlinear relationship, which is not possible through classical statistical methodologies except for mechanical curve fitting procedures, which minimize the sum of the squared errors without explicit logical foundation. However, their treatment by fuzzy sets is possible.

5.6.1 LOGICAL PROPORTIONAL RELATION

Now it is time to indicate the simple way of obtaining various lines or curves on the basis of fuzzy matrix relationship as two inputs (I_A, I_B) and a single output (O) linguistic variable. Both $I_A = [I_{A1}, I_{A2}, I_{A3}, I_{A4}]$, $I_B = [I_{B1}, I_{B2}, I_{B3}]$, and $O = [O_{31}, O_{22}, O_{13}, O_{14}]$ are valid fuzzy subsets. Herein, the I_A and I_B linguistic input variables have four and three sets (crisp or fuzzy), respectively, and hence the output space has $3 \times 4 = 12$ mutually exclusive or inclusive sub-areas, as shown in Table 5.3. The output space as defined in this table shows a different pattern as presented in Figures 5.11

TABLE 5.3

Proportional Relation Matrix

		Input, I_A			
		I_{A1}	I_{A2}	I_{A3}	I_{A4}
Input, I_B	I_{B1}			O_{13}	O_{14}
	I_{B2}		O_{22}		
	I_{B3}	O_{31}			

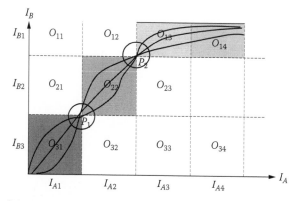

FIGURE 5.11 Crisp system and direct relationship curves.

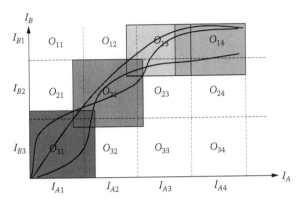

FIGURE 5.12 Fuzzy system and direct relationship curves.

and 5.12 for crisp and fuzzy sub-areas, respectively. In Figure 5.11, a bundle of various possible curves are drawn and there are many other alternatives as the reader can appreciate. Such curves are possibilities without any data availability. Herein, P_1 and P_2 are crisp transition locations between sub-areas (subsets). However, in the case of fuzzy relations, the transition points leave their positions to overlapping transition areas, which make the variability domain of the relationship wider than in Figure 5.12.

5.6.2 LOGICAL INVERSE RELATION

Similar to the previous case of proportional relationships, it is possible to visualize the inversely proportional case as shown in Table 5.4, where only four output sub-areas are possible. The appropriate crisp and fuzzy Cartesian coordinate system indications corresponding to this table are given in Figures 5.13 and 5.14, respectively. It is possible to trace an infinite number of continuous or noncontinuous curves on the available subsets but on the condition that the transitions are through crisp points P_1 and P_2. In Figure 5.13, the transition points leave their position to overlapping transitional areas, which makes the variability domain of the relationship wider than in Figure 5.14.

TABLE 5.4
Inverse Relation Matrix

		Output, I_A			
		I_{A1}	I_{A2}	I_{A3}	I_{A4}
Input, I_B	I_{B1}	O_{11}			
	I_{B2}		O_{22}		
	I_{B3}			O_{33}	O_{34}

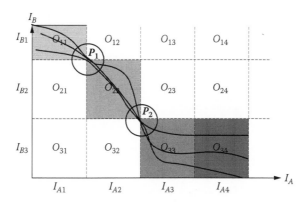

FIGURE 5.13 Crisp system and indirect relationship curves.

5.6.3 LOGICAL HAPHAZARD RELATION

It is well known from statistical studies in hydrology that there are relationships or categories of data that can be described neither mathematically nor statistically. However, provided that the partial relationships (rules, see Chapter 6) between the fuzzy subsets are identifiable, one can visualize the pattern on the Cartesian coordinate space. For instance, the relationship matrix might have no consistency between input and output variables, as in Table 5.5. Figure 5.15 indicates the crisp patterns on the Cartesian coordinate space. Herein, the mosaic of valid sub-areas is rather random, without definite curves or lines. However, such a representation provides categories that are not in harmony (relationship) with each other. Such haphazard sub-areas help define valid fuzzy rules (see Chapter 6). It is obvious that the subsets are mutually exclusive but not exhaustive over the output variation domain. In a complete random relationship, all the sub-areas are mutually exclusive and exhaustive, as in the mathematical studies. However, in Figure 5.16, the corresponding fuzzy relationship pattern is presented. In this figure, haphazard sub-areas are neither mutually exclusive nor exhaustive.

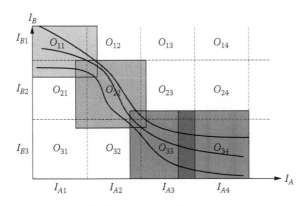

FIGURE 5.14 Fuzzy system and indirect relationship curves.

TABLE 5.5
Haphazard Relation Matrix

		Input, I_A			
		I_{A1}	I_{A2}	I_{A3}	I_{A4}
Input, I_B	I_{B1}		O_{12}		O_{14}
	I_{B2}	O_{21}			O_{24}
	I_{B3}		O_{32}	O_{33}	

5.6.4 Logical Extreme Cases

Some extreme hydrological cases can be shown and expressed in terms of fuzzy subsets. In such extreme cases, the sub-areas are distinct from each other, either in the crisp or fuzzy sense. Table 5.6 indicates a relationship matrix with three extreme cases. There are three valid sub-areas and hence only the corresponding crisp and fuzzy patterns are given in Figures 5.17 and 5.18, respectively. In Figure 5.18, the variation domain of each sub-area becomes wider but in most cases is completely intact.

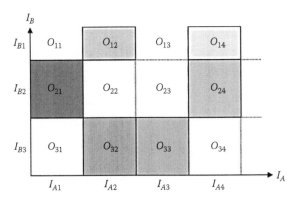

FIGURE 5.15 Crisp system and haphazard sub-areas.

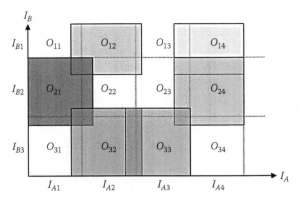

FIGURE 5.16 Fuzzy system and haphazard areas.

Table 5.6
Extreme Area Partition

		Input, I_A			
		I_{A1}	I_{A2}	I_{A3}	I_{A4}
Input, I_B	I_{B1}				O_{14}
	I_{B2}	O_{21}			
	I_{B3}			O_{33}	

5.6.5 Climate Classification

The two most important variables in any climate classification are the temperature and the amount of rainfall. According to these two meteorological variables, Table 5.7 is prepared for the climate classification with fuzzy words. The crisp and fuzzy representations of this table are presented in Figures 5.19 and 5.20, respectively.

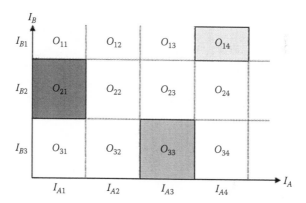

FIGURE 5.17 Crisp system and extreme sub-areas.

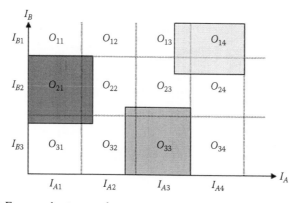

FIGURE 5.18 Fuzzy and extreme sub-areas.

TABLE 5.7
Fuzzy Climate Classification

		'Temperature'			
		"cold"	"warm"	"medium"	"hot"
"Rainfall"	"light"	"polar"	"semi-arid"	"arid"	"E. Arid"
	"moderate"	"humid"	"temperate"	"steppe"	"E. steppe"
	"heavy"	"E. humid"	"semi-humid"	"subtropical"	"tropical"

The words can either be subdivided descriptively or in many cases they are also related to some intuitive, daily, or expert numbers. Depending on the fuzzy subsets as one from temperature and the other from rainfall, it is possible to allocate the climate type for a region. Of course, the numerical values for temperature and rainfall fuzzy sets depend on the location of the study area. The strength of climate can be visualized in each sub-area by considering that the centroid of the sub-area has maximum MD equal to 1.

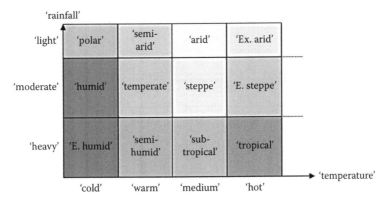

FIGURE 5.19 Crisp Cartesian coordinate system and climate classification.

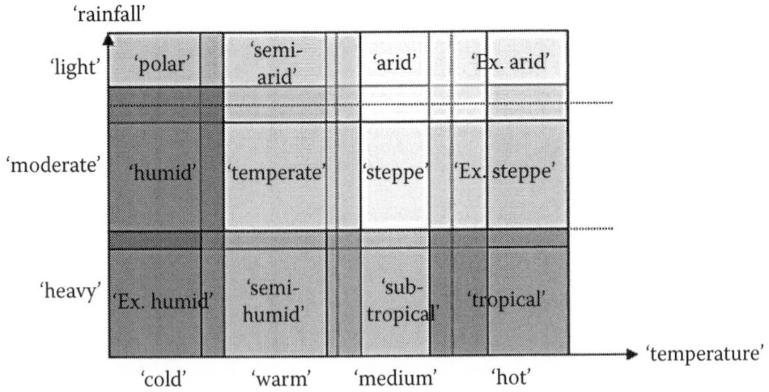

FIGURE 5.20 Fuzzy system and climate classification.

5.7 FUZZY CLUSTERING ALGORITHMS

Categorization is the basic human classification of different but similar objects into the same group so as to recognize them in a general manner (see Chapter 2). In Aristotelian categorization (i.e., CL), there are two classes that are mutually exclusive and exhaustive. It was previously explained that in a fuzzy classification, mutually exclusiveness and exhaustiveness concepts are not valid because there are categories that are partially included within each other. Hence, different categorizations may have overlapping sectors at different degrees, and they are still regarded as distinct from each other. For instance, humans grasp what rivers are; but if a direct question is asked, such as distinction between many rivers, perhaps one might be inclined to regard the "longest" ones by FL as rivers and others "not rivers" by CL where humans are forced to make such crisp categorizations. This is contrary to what humans are bound to think naturally because they perceive and think in terms of the granulation of objects. Automatically, in any language, the categorization starts at least as "low," "medium," and "high" rivers, in contrast to classical two-way logic, such as "to be" or "not to be."

Clustering can be considered the most important *unsupervised learning* problem; it deals with finding a *structure* in a collection of unlabeled data.

5.7.1 Distance Measure

An important component of any clustering algorithm is the distance measure between data points on any scatter diagram. If the components of the data vectors are all in the same physical units, then it is possible that the simple Euclidean distance metric is sufficient to successfully group similar data instances. However, even in this case, the Euclidean distance can sometimes be misleading. Figure 5.21 illustrates this with an example of the width and height measurements of an object. Despite both measurements being taken in the same physical units, an informed decision must be made as to the relative scaling. As is obvious, different scaling can lead to different clustering.

Notice that this is not only a graphic issue. The problem arises from the mathematical formula used to combine the distances between the single components of the

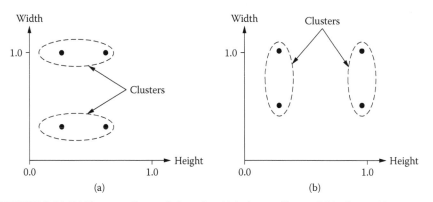

FIGURE 5.21 Different scaling and clustering (a) before scaling and (b) after scaling.

data feature vectors into a unique distance measure that can be used for clustering purposes. Different formulae lead to different clusterings. Distance-based categorization was discussed in Chapter 3 for the attachment of MFs.

Again, domain knowledge must be used to guide the formulation of a suitable distance measure for each particular application. For higher dimensional data, a popular measure is the Minkowski metric:

$$d_p(x_i, x_j) = \left(\sum_{k=1}^{n} |x_{i,k} - x_{j,k}|^p \right)^{\frac{1}{p}} \tag{5.13}$$

where d is the dimensionality of the data. The Euclidean distance (Equation 3.26 in Chapter 3) is a special case where $p = 2$, while the Manhattan metric has $p = 1$. However, there are no general theoretical guidelines for selecting a measure for any given application.

It is often the case that the components of the data feature vectors are not immediately comparable. It may be that the components are not continuous variables, such as length, but rather nominal scales (see Chapter 2), such as days of the week.

5.7.2 *κ*-Means

This method is one of the simplest unsupervised learning algorithms that solves the well-known exclusive clustering problem (MacQueen, 1967). The procedure follows a simple and easy way to classify a given data set through a certain number of clusters (assume k clusters) fixed *a priori*. The main idea is to define k centroids, one for each cluster. These centroids should be placed in a cunning way because different locations cause different results. So, the better choice is to place them, as much as possible, far away from each other. The next step is to take each point belonging to a given data set and associate it to the nearest centroid. When no point is pending, the first step is completed and an early grouping is done. At this point, one needs to recalculate k new centroids by taking the arithmetic averages of the data values in each group as new centers of the clusters resulting from the previous step. After one has these k new centroids, a new binding must be done between the same data-set points and the nearest new centroid. Hence, a loop has been generated. One may notice that the k centroids change their location step by step until no more changes are necessary. That is, centroids do not move anymore. Finally, this algorithm aims to minimize an objective function, which is given as:

$$J = \sum_{j=1}^{k} \sum_{i=1}^{n} \left\| x_i^{(j)} - c_j \right\|^2 \tag{5.14}$$

where $|x_i^{(j)} - c_j|^2$ is a chosen distance measure between a data point $x_i^{(j)}$, and the cluster center c_j is an indicator of the distance of n data points from their respective cluster centers. Suppose that there are n sample feature vectors $\mathbf{x}_1, \mathbf{x}_2, \ldots, \mathbf{x}_n$ all from

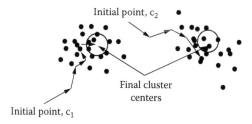

Initial point, c_2

Final cluster
centers

Initial point, c_1

FIGURE 5.22 Means moving toward cluster centers.

the same class, and one knows that they fall into k compact clusters, $k < n$. Let c_j be the initial mean of the vectors in cluster j. If the clusters are well separated, one can use a minimum-distance classifier to separate them, where one can say that x is in cluster j if $\|x - c_j\|$ is the minimum of all k distances. This suggests the following procedure for finding the k-means:

1. Make initial guesses for the means (cluster centers) $c_1, c_2, \ldots, c_k$,
2. Until there are no changes in successive means, use the estimated means to classify the samples into clusters for $j = 1, 2, \ldots, k$, replace c_j with the mean of all the samples for cluster j, and continue until the last data point is executed.

Figure 5.22 gives an example in two-dimensional space to show how the means c_1 and c_2 move into the centers of two clusters.

This is a simple version of the k-means procedure. It can be viewed as a greedy algorithm for partitioning the n samples into k clusters so as to minimize the sum of the squared distances to the cluster centers. It does, however, have the following weaknesses:

1. The way to initialize the means is not specified. One popular way to start is to randomly choose k of the center samples.
2. The results produced depend on the initial values for the means, and it frequently happens that sub-optimal partitions are found. The standard solution is to try a number of different starting points.
3. It may happen that the set of samples closest to c_j is empty, so that c_j cannot be updated. This is an annoyance that must be handled in an implementation.
4. The results depend on the metric used to measure $\|x - c_j\|$. A popular solution is to normalize each variable by its standard deviation, although this is not always desirable.

The results depend on the value of k. This problem is particularly troublesome because one often has no way of knowing how many clusters exist. Unfortunately, there is no general theoretical solution to find the optimal number of clusters for any given data set. A simple approach is to compare the results of multiple runs with different k classes and choose the best one according to a given criterion, but one must be careful because increasing k results in smaller error function values by definition, and also an increasing risk of overfitting.

Example 5.12

To apply the k-means algorithm, the monthly sunshine duration ratio S/S_0 and solar irradiation ratio H/H_0 data are given in Table 5.8 for the southern part of Turkey for the Mediterranean province city, Adana. Herein. S and H are terrestrial sunshine duration and solar irradiation measurements, whereas S_0 and H_0 are the extraterrestrial (astronomical) counterparts. Because always $S < S_0$ and $H < H_0$, the aforementioned ratios have variation domains between 0 and 1.

Because there are four seasons in this part of the world (i.e., Turkey), the number of clusters is taken as four and hence four initial cluster center coordinates

TABLE 5.8
Solar Energy Data

S/S_0	H/H_0	S/S_0	H/H_0	S/S_0	H/H_0	S/S_0	H/H_0
0.460	0.397	0.633	0.577	0.773	0.531	0.572	0.537
0.513	0.434	0.464	0.578	0.799	0.620	0.695	0.576
0.479	0.469	0.737	0.572	0.731	0.569	0.705	0.569
0.427	0.523	0.240	0.537	0.606	0.512	0.593	0.531
0.630	0.570	0.404	0.511	0.639	0.469	0.799	0.615
0.783	0.535	0.534	0.468	0.430	0.650	0.695	0.580
0.751	0.554	0.511	0.445	0.480	0.430	0.673	0.567
0.784	0.516	0.543	0.452	0.387	0.392	0.590	0.530
0.721	0.509	0.726	0.493	0.370	0.407	0.613	0.606
0.690	0.494	0.839	0.521	0.502	0.448	0.561	0.486
0.645	0.509	0.776	0.484	0.617	0.480	0.416	0.439
0.711	0.450	0.731	0.495	0.673	0.483	0.442	0.469
0.486	0.372	0.679	0.492	0.832	0.557	0.492	0.498
0.428	0.368	0.539	0.432	0.792	0.523	0.622	0.541
0.486	0.424	0.571	0.543	0.716	0.510	0.733	0.571
0.487	0.468	0.480	0.408	0.594	0.409	0.689	0.560
0.599	0.420	0.490	0.406	0.477	0.447	0.719	0.543
0.712	0.453	0.555	0.446	0.481	0.533	0.695	0.580
0.702	0.541	0.651	0.471	0.694	0.529	0.577	0.488
0.822	0.546	0.534	0.403	0.621	0.504	0.571	0.479
0.737	0.578	0.673	0.492	0.529	0.486	0.613	0.606
0.617	0.583	0.702	0.509	0.686	0.529	0.555	0.486
0.667	0.415	0.769	0.513	0.753	0.591	0.416	0.456
0.754	0.610	0.705	0.533	0.660	0.559	0.463	0.491
0.428	0.428	0.617	0.517	0.751	0.601	0.595	0.543
0.530	0.482	0.652	0.514	0.643	0.554	0.530	0.502
0.381	0.422	0.728	0.548	0.705	0.573	0.523	0.504
0.506	0.463	0.340	0.334	0.611	0.555	0.766	0.601
0.704	0.518	0.519	0.435	0.564	0.481	0.784	0.597
0.733	0.537	0.458	0.439	0.613	0.614	0.655	0.566
0.766	0.535	0.572	0.465	0.580	0.540	0.629	0.539

TABLE 5.9
Cluster Center Coordinates

Center	Sunshine Duration Ratio	Solar Irradiation Ratio	Seasons
C1	0.4635	0.4556	Autumn
C2	0.6245	0.4758	Winter
C3	0.6164	0.5618	Spring
C4	0.7457	0.5502	Summer

are selected randomly. After execution of the above k-means steps, the center coordinates are those presented in Table 5.9. In the same table, meaningful seasons are also provided in the rightmost column. The plot of each cluster data with a different symbol and the corresponding centers in small circles are shown in Figure 5.23. Each one of the centers corresponds to a specific season, as indicated in the rightmost column of Table 5.9. In addition, Table 5.10 provides the data belonging to a specific cluster in a crisp manner. It is obvious that, herein, each data point fully belongs to one of the clusters (C1, C2, C3, or C4) and there is no partial belongingness, which is possible in the case of the fuzzy c-means clustering technique.

5.7.3 C-Means

In the literature, fuzzy c-means (FCM) has been reported as being one of the most promising clustering methods (Bezdek, 1981; Dunn, 1973). It is an iterative clustering approach that resembles the well-known k-means technique but it uses fuzzy

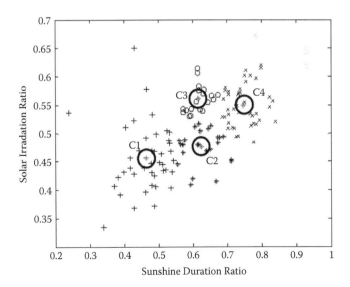

FIGURE 5.23 Data and cluster centers.

TABLE 5.10
Cluster Numbers for Each Data Point

Cluster No.	S/S_0	H/H_0	Cluster No.	S/S_0	H/H_0	Cluster No.	S/S_0	H/H_0	Cluster No.	S/S_0	H/H_0
C1	0.63	0.57	C2	0.46	0.397	C3	0.783	0.535	C4	0.69	0.494
C1	0.617	0.583	C2	0.513	0.434	C3	0.751	0.554	C4	0.645	0.509
C1	0.633	0.577	C2	0.479	0.469	C3	0.784	0.516	C4	0.711	0.45
C1	0.571	0.543	C2	0.427	0.523	C3	0.721	0.509	C4	0.599	0.42
C1	0.66	0.559	C2	0.486	0.372	C3	0.702	0.541	C4	0.712	0.453
C1	0.643	0.554	C2	0.428	0.368	C3	0.822	0.546	C4	0.667	0.415
C1	0.611	0.555	C2	0.486	0.424	C3	0.737	0.578	C4	0.679	0.492
C1	0.613	0.614	C2	0.487	0.468	C3	0.754	0.61	C4	0.555	0.446
C1	0.58	0.54	C2	0.428	0.428	C3	0.704	0.518	C4	0.651	0.471
C1	0.617	0.575	C2	0.53	0.482	C3	0.733	0.537	C4	0.673	0.492
C1	0.572	0.537	C2	0.381	0.422	C3	0.766	0.535	C4	0.617	0.517
C1	0.593	0.531	C2	0.506	0.463	C3	0.822	0.564	C4	0.652	0.514
C1	0.673	0.567	C2	0.464	0.578	C3	0.726	0.552	C4	0.572	0.465
C1	0.59	0.53	C2	0.24	0.537	C3	0.737	0.572	C4	0.613	0.479
C1	0.613	0.606	C2	0.404	0.511	C3	0.726	0.493	C4	0.606	0.512
C1	0.622	0.541	C2	0.534	0.468	C3	0.839	0.521	C4	0.639	0.469
C1	0.613	0.606	C2	0.511	0.445	C3	0.776	0.484	C4	0.617	0.48
C1	0.595	0.543	C2	0.543	0.452	C3	0.731	0.495	C4	0.673	0.483
C1	0.655	0.566	C2	0.539	0.432	C3	0.702	0.509	C4	0.594	0.409
C1	0.629	0.539	C2	0.48	0.408	C3	0.769	0.513	C4	0.621	0.504
			C2	0.49	0.406	C3	0.705	0.533	C4	0.564	0.481
			C2	0.534	0.403	C3	0.728	0.548	C4	0.561	0.486
			C2	0.34	0.334	C3	0.776	0.517	C4	0.577	0.488
			C2	0.519	0.435	C3	0.773	0.531	C4	0.571	0.479
			C2	0.458	0.439	C3	0.799	0.62	C4	0.555	0.486
			C2	0.43	0.65	C3	0.731	0.569			
			C2	0.48	0.43	C3	0.832	0.557			
			C2	0.387	0.392	C3	0.792	0.523			
			C2	0.37	0.407	C3	0.716	0.51			
			C2	0.502	0.448	C3	0.694	0.529			
			C2	0.477	0.447	C3	0.686	0.529			
			C2	0.481	0.533	C3	0.753	0.591			
			C2	0.529	0.486	C3	0.751	0.601			
			C2	0.399	0.432	C3	0.705	0.573			
			C2	0.416	0.439	C3	0.695	0.576			
			C2	0.442	0.469	C3	0.705	0.569			
			C2	0.492	0.498	C3	0.799	0.615			
			C2	0.416	0.456	C3	0.695	0.58			
			C2	0.463	0.491	C3	0.733	0.571			
			C2	0.53	0.502	C3	0.689	0.56			
			C2	0.523	0.504	C3	0.719	0.543			
						C3	0.695	0.58			
						C3	0.766	0.601			
						C3	0.784	0.597			

MDs instead of crisp cluster numbers. FCM partitions the data set $X = x_1, x_2, \ldots,$ x_n into c fuzzy subsets with $u_i(x_k)$ MD of data x_k in class i. The values of $u_i(x_k)$ are arranged as a $c \times n$ matrix, U. The method approximately minimizes the sum of the squared error (objective) function defined as:

$$J_m(U,C:X) = \sum_{i=1}^{c} \sum_{k=1}^{n} (u_{ik})^m \, \| x_k - c_i \| \tag{5.15}$$

where $C = [c_1, c_2, \ldots, c_c]$ is a set of cluster centers and $m > 1$ is a weighting exponent affecting the fuzziness of U. The parameters (U, C) may minimize J_m only if u_{ik} and c_i are defined as (Ross, 1995):

$$u_{ik} = \left[\sum_{j=1}^{c} \left(\frac{\| x_k - c_i \|}{\| x_k - c_j \|} \right)^{2/(m-1)} \right]^{-1} \quad \text{for all } i, k \tag{5.16}$$

and

$$c_i = \frac{\displaystyle\sum_{k=1}^{n} (u_{ik})^m x_k}{\displaystyle\sum_{k=1}^{n} (u_{ik})^m} \quad \text{for all } i \tag{5.17}$$

The stop criteria for the method is determined by $E_t \leq \varepsilon$, where

$$E_t = \sum_{i=1}^{c} \| c_{i,t+1} - c_{i,t} \| \quad \text{for all } t \tag{5.18}$$

The method contains the following iteration steps:

1. Initialize matrix U_0.
2. Choose c, m, and ε.
3. Compute all c cluster centers $c_{i,0}$.
4. Compute all cxn memberships $u_{ik,\,t}$ and update all c cluster centers $c_{i,\,t+1}$.
5. Compute E_t.
5. If $E_t \leq \varepsilon$, then stop; else return to 3.

When the stop criteria are fulfilled, a fuzzy output is available. However, in further processing, a crisp valued output is necessary. To produce a crisp valued output of the fuzzy c-means algorithm, maximum MDs are found; that is, if max (u_{ik}) = u_{1k}, then x_k is assigned the label associated with class 1. The flowchart for the FCM algorithm is shown in Figure 5.24.

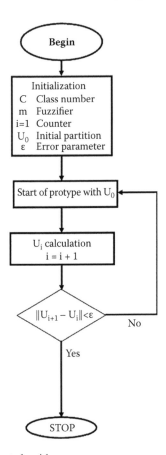

FIGURE 5.24 FCM flowchart algorithm.

Example 5.13

The same data as in Example 5.12 are used for FCM application and the final results with four cluster centers (C1, C2, C3, and C4) are presented in Table 5.11. On the other hand, the MDs of each data point in four clusters are given in the last four columns of Table 5.12 depending on seasons. It is obvious that the summation of each data point MDs in each cluster is equal to 1, as required by the FCM procedure. However, by taking the maximum MD for each data point, its crisp

TABLE 5.11

FCM Cluster Centers

Center	Sunshine Duration Ratio	Solar Irradiation Ratio
C1	0.4241	0.4348
C2	0.5322	0.4681
C3	0.6355	0.5374
C4	0.7502	0.5488

TABLE 5.12
FCM Data MDs

S/So	*H/Ho*	C1	C2	C3	C4
0.460	0.397	0.200	0.019	0.041	0.740
0.513	0.434	0.782	0.017	0.047	0.154
0.479	0.469	0.548	0.020	0.054	0.379
0.427	0.523	0.304	0.041	0.098	0.557
0.630	0.570	0.048	0.065	0.872	0.016
0.783	0.535	0.017	0.921	0.054	0.008
0.751	0.554	0.001	0.997	0.002	0.000
0.784	0.516	0.029	0.872	0.085	0.014
0.721	0.509	0.047	0.721	0.215	0.019
0.690	0.494	0.095	0.370	0.502	0.033
0.645	0.509	0.056	0.064	0.866	0.015
0.711	0.450	0.150	0.429	0.362	0.059
0.486	0.372	0.357	0.040	0.082	0.521
0.428	0.368	0.163	0.025	0.047	0.765
0.486	0.424	0.459	0.022	0.053	0.466
0.487	0.468	0.667	0.018	0.051	0.265
0.599	0.420	0.541	0.094	0.245	0.120
0.712	0.453	0.144	0.440	0.360	0.057
0.702	0.541	0.043	0.617	0.325	0.016
0.822	0.546	0.047	0.808	0.121	0.025
0.737	0.578	0.017	0.896	0.079	0.008
0.617	0.583	0.092	0.099	0.777	0.032
0.667	0.415	0.285	0.241	0.374	0.101
0.754	0.610	0.043	0.782	0.154	0.021
0.428	0.428	0.005	0.001	0.001	0.994
0.530	0.482	0.969	0.004	0.013	0.014
0.381	0.422	0.073	0.012	0.023	0.892
0.506	0.463	0.882	0.009	0.028	0.082
0.704	0.518	0.055	0.578	0.346	0.021
0.733	0.537	0.009	0.946	0.042	0.004
0.766	0.535	0.007	0.965	0.025	0.003
0.822	0.564	0.047	0.805	0.123	0.025
0.726	0.552	0.012	0.919	0.064	0.005
0.633	0.577	0.060	0.089	0.830	0.021
0.464	0.578	0.401	0.081	0.216	0.303
0.737	0.572	0.013	0.924	0.058	0.006
0.240	0.537	0.253	0.088	0.146	0.514
0.404	0.511	0.227	0.034	0.076	0.663
0.534	0.468	0.999	0.000	0.000	0.000
0.511	0.445	0.845	0.012	0.035	0.108

TABLE 5.13
Data MDs

Data Number	X Values	Y Values
1	0.2190	0.5297
2	0.0470	0.6711
3	0.6789	0.0077
4	0.6793	0.3834
5	0.9347	0.0668
6	0.3835	0.4175
7	0.5194	0.6868
8	0.8310	0.5890
9	0.0346	0.9304
10	0.0535	0.8462

group can be identified among the four clusters. For instance, the first data point in Table 5.13 belongs crisply to cluster C4. Figure 5.25 indicates the scatter of data with different symbols and cluster centers are shown by circles.

The FCM also provides a basis for the decomposition of a data set according to MDs into four descriptors as shown in Figure 5.26. For instance, Figure 5.26a shows the variation of data MDs within C1. It is also interesting to look at the variation of the objective (error) function according to Equation (5.15) as shown in Figure 5.27. It is obvious that even though the initial four center coordinates are chosen randomly, after five to six iterations, the objective function attains a stable value according to the criterion in Equation (5.18) and hence the iteration is stopped.

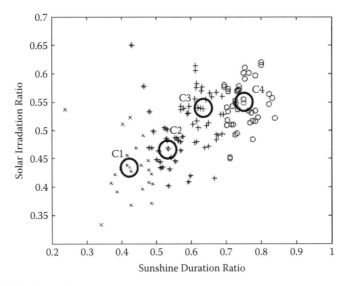

FIGURE 5.25 Fuzzy cluster centers.

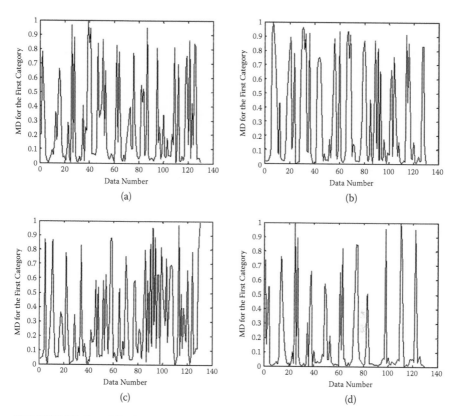

FIGURE 5.26 Data MD decomposition into four groups.

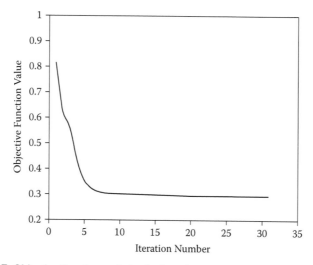

FIGURE 5.27 Objective function variation by iteration.

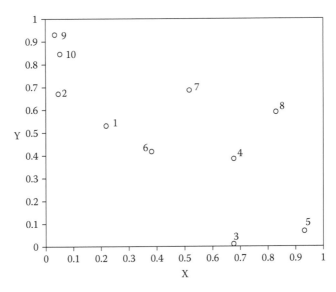

FIGURE 5.28 Data scatter points.

TABLE 5.14
Fuzzy Cluster Center Coordinates

Center	X	Y
C1	0.7349	0.2821
C2	0.1449	0.7231

TABLE 5.15
Data MDs

Data Number	Fuzzy Partition	
	C1	C2
1	0.11	0.89
2	0.01	0.99
3	0.91	0.09
4	0.97	0.03
5	0.92	0.08
6	0.51	0.49
7	0.40	0.60
8	0.82	0.18
9	0.05	0.95
10	0.02	0.98

Example 5.14

Another FCM application is performed for the data given in Table 5.13 where there are only ten values. Finally, the scatter of these data is given in Figure 5.28. The application of the FCM algorithm with two cluster centers leads to the results in Table 5.14. Accordingly, MDs for each data are presented in Table 5.15. Again it is obvious that the MDs summation is equal to 1. The variation of the objective function is given in Table 5.16. Finally, Figure 5.29 indicates the distribution of the MDs on scatter diagrams for each cluster.

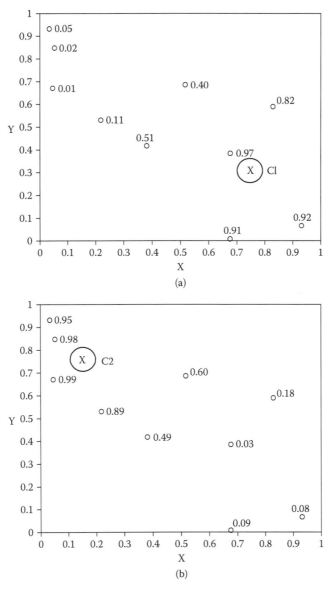

FIGURE 5.29 Data MD distribution for (a) C1 and (b) C2.

TABLE 5.16
Objective Function Values

Iteration	Objective Function Value
1	1.195751
2	0.895094
3	0.753837
4	0.568045
5	0.533401
6	0.531844
7	0.531776
8	0.531772

REFERENCES

Bezdek, J.C. 1981. *Pattern Recognition with Fuzzy Objective Function Algorithms.* Plenum Press, New York.

Dunn, J.C. 1973. A fuzzy relative of the ISODATA process and its use in detecting compact well-separated clusters. *J. Cybernet.* 3: 32–57.

MacQueen, J.B. 1967. Some Methods for Classification and Analysis of Multivariate Observations. *Proceedings of 5th Berkeley Symposium on Mathematical Statistics and Probability,* Berkeley, University of California Press, 1: 281–29.

Ross, J.T. 1995. *Fuzzy Logic with Engineering Applications.* McGraw-Hill, New York, 593 pp.

Şen, Z. 1978. Autorun analysis of hydrologic time-series. *J. Hydrology* 36 (1-2): 75–85.

Şen, Z. 1979. Application of the autorun test to hydrologic data. *Journal of Hydrology* 42(1–2): 1–7.

Şen, Z. 2001. *Bulanık Mantık ve Modelleme İlkeleri (Fuzzy Logic and Modelling Principles),* Bilgi, Sanat ve Kültür Basimevi, İstanbul (in Turkish).

PROBLEMS

5.1 Map the following two sets of words mutually:
   (a) Force ("weak," "medium," "strong")
   (b) Acceleration ("high," "medium," "low")

5.2 Relate the words that describe the following two phenomena:
   (a) Rainfall ("low," "medium," "intense")
   (b) Runoff ("weak," "medium," "strong")

5.3 If the nonlinear relationship between "x" and "y" is in the form of a simple parabola as:

$$(``y") = (``x")^2$$

then after drawing this fuzzy relationship, roughly write the relationship linguistically by considering five fuzzy subsets for each linguistic variable.

5.4 Show on the Cartesian coordinate system the following set of relationships and then make your interpretation.

"IF *X* is "high" THEN *Y* is 'low.'"

"IF *X* is "high" THEN *Y* is 'high.'"

"IF *X* is "medium" THEN *Y* is 'medium.'"

"IF *X* is "low" THEN *Y* is 'low.'"

"IF *X* is "low" THEN *Y* is 'high.'"

5.5 Write the linguistic relationship between "*U*" and "*V*" linguistic variables from the following graph by attaching fuzzy sets for each variable.

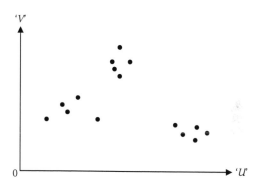

5.6 Interpret the scatter of points in the following graphs with appropriate linguistic variables.

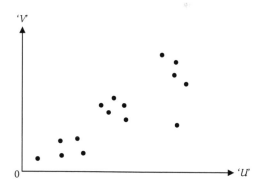

5.7 Interpret the differences between the two scatter diagrams given below from statistical and fuzzy logic points of view.

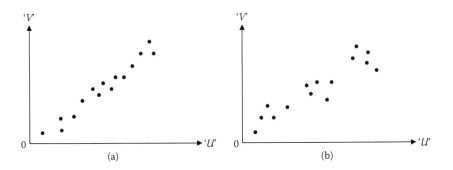

(a)   (b)

5.8 The transition matrix elements between yesterday and today, "YT," and today and tomorrow, "TT," are given as follows:

$$
"YT" = \begin{bmatrix}
\dfrac{0.1}{5.1} & \dfrac{0.3}{8.3} & \dfrac{0.5}{11.3} & \dfrac{1.0}{13.1} \\[2ex]
\dfrac{0.3}{3.2} & \dfrac{0.5}{6.1} & \dfrac{0.7}{9.2} & \dfrac{0.9}{11.7} \\[2ex]
\dfrac{0.5}{2.4} & \dfrac{0.7}{4.5} & \dfrac{0.9}{7.6} & \dfrac{1.0}{8.8} \\[2ex]
\dfrac{0.7}{1.9} & \dfrac{0.8}{3.4} & \dfrac{0.9}{6.1} & \dfrac{1.0}{5.4}
\end{bmatrix}
\quad \text{and} \quad
"TT" = \begin{bmatrix}
\dfrac{0.2}{5.9} & \dfrac{0.5}{7.3} & \dfrac{0.7}{10.2} & \dfrac{1.0}{11.0} \\[2ex]
\dfrac{0.1}{2.9} & \dfrac{0.3}{5.8} & \dfrac{0.7}{8.1} & \dfrac{1.0}{9.9} \\[2ex]
\dfrac{0.4}{1.8} & \dfrac{0.6}{3.7} & \dfrac{0.8}{8.7} & \dfrac{1.0}{9.6} \\[2ex]
\dfrac{0.5}{2.1} & \dfrac{0.7}{3.3} & \dfrac{0.8}{5.7} & \dfrac{1.0}{4.3}
\end{bmatrix}
$$

Calculate the relationship between the transition from yesterday to tomorrow, "YT."

5.9 Plot the following fuzzy numbers on a Cartesian coordinate system by showing the MDs of each point on this scatter diagram.

$$
"X" = \left\{ \frac{0.5}{3} + \frac{0.7}{4} + \frac{0.9}{7} + \frac{0.96}{10} + \frac{1.0}{11} \right\}
$$

$$
"Y" = \left\{ \frac{0.2}{5.1} + \frac{0.4}{8.3} + \frac{0.6}{9.7} + \frac{1.0}{12.3} + \frac{0.8}{15.1} \right\}
$$

5.10 Fill the convenient overlapping sub-areas in the following graph for successive daily flow records such that one gets the impression that "high" flows follow "low" flows; "low" flows follow "high" flows; and "middle flows" follow "high" flows.

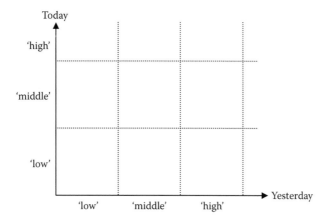

5.11 If the FAM for daily rainfall measurements between two successive days (D) are given on the basis of five fuzzy sets as "very low" ("VL"), "low" ("L"), "medium" ("M"), "high" ("H"), and "very high" ("VH") as follows:

$$R(``D",``D") = \begin{bmatrix} 1.0 & 0.0 & 0.7 & 0.6 & 0.8 \\ 0.0 & 1.0 & 0.6 & 0.2 & 0.5 \\ 0.7 & 0.6 & 1.0 & 0.1 & 0.3 \\ 0.6 & 0.2 & 0.5 & 1.0 & 1.0 \\ 0.8 & 0.5 & 0.3 & 1.0 & 1.0 \end{bmatrix}$$

find the following fuzzy relationships:
(a) FAM between today and after tomorrow
(b) FAM between today and 2 days ago
(c) FAM between 3 January and 5 January

5.12 There are four graphs of different water qualities among which a fuzzy relationship exists given the following matrix:

$$W = \begin{bmatrix} 1 & 0.9 & 0.4 & 0.6 \\ 0.9 & 1.0 & 0.8 & 0.1 \\ 0.4 & 0.8 & 1.0 & 0.5 \\ 0.6 & 0.1 & 0.5 & 1.0 \end{bmatrix}$$

(a) Which groups are similar at the 0.4 α-cut level?
(b) What about at the 0.8 α-cut level?

5.13 A hydrologist evaluates five catchments by their three features: water potential, agriculture, and climate. He or she becomes an expert with fuzzy concepts concerning the features for the catchments. If he or she uses an α-cut level of 0.8, then how many classes can he or she make out of these catchments?

$$
R(C,C) = \begin{bmatrix}
1.0 & 0.2 & 0.4 & 0.9 & 0.6 \\
0.2 & 1.0 & 0.5 & 0.3 & 0.7 \\
0.5 & 0.6 & 1.0 & 0.4 & 0.3 \\
0.9 & 0.4 & 0.3 & 1.0 & 0.5 \\
0.6 & 0.7 & 0.3 & 0.5 & 1.0
\end{bmatrix}
$$

6 Fuzzy Logical Rules

6.1 GENERAL

The generating mechanism of any hydrologic event is a harmonious set of relationships between the input and output variables, which can be represented by different methodologies such as mathematics, probability, statistics, and stochastic processes, all of which have rational and logical foundations. It is necessary to have mathematical formulations for such methodologies and they are all based on CL as dependent or independent variables. If dependent, then again a CL relationship in the form of direct or indirect proportionality statements takes place in the construction of a suitable model. In the CL model, a set of assumptions, simplification statements, and parameters are necessary.

Model construction expansion of CL into FL necessitates detailed logical relationships among the input and output variables based on fuzzy sets without restrictive assumptions and parameters. The generating mechanism of a hydrological phenomenon is expressed in a set of rules, which constitutes the rule base and prior to any database, such a valid rule base is the fundamental requirement in FL modeling. Hence, a rule base is the collection of all possible rules, each with an explanation of partial behaviors, that cover the overall behavior of the hydrological event. There are different approaches in the establishment of a valid rule base, including mechanical, personal logical inspirations, and expert views about hydrologic events and database usage. However, prior to such a rule base, fuzzification of all the input and output linguistic variables is necessary, and the rule base is trained finally by the database, if available. Fuzzifications include greater generality, higher expressive power, an enhanced ability to model real-world problems, and a methodology for exploiting the tolerance for imprecision. Hence, FL can help achieve practicality, reboustness, and lower solution cost. Because each rule explains partial behavior of the event, it is not possible to have the validity of all the rules in the rule base for a given input data. Hence, any given input data validates few rules and in this case these rules are said to be *triggered* or *fired*.

The primary purpose of this chapter is to provide the necessary steps in rule base construction prior to a complete modeling by explaining fuzzification, rule base establishment, and triggering stages.

6.2 FUZZIFICATION

The purpose of modeling is to map the input variables onto output in such a way that the mapping provides output variables with a minimum possible error. In FL modeling, it is not necessary to have a database at the beginning. The first step is to identify the input and output linguistic variables and then decompose them into formal fuzzy subsets. In the fuzzification stage, the following points are important:

1. In hydrological modeling, there is a single output variable, whereas there may be many input variables. In general, it is multiple-input-single-output (MISO) modeling.
2. The variability range of each linguistic variable must be determined linguistically as closed-interval, left-open-interval, right-open-interval, or open-interval from both sides. For instance, the "porosity" linguistic variable has closed-interval because it is confined between 0 and 1, inclusive; "rainfall" has right-open interval; "water deficit" should have left-open interval; and "temperature" has open-interval,
3. The number of fuzzy sets for linguistic variable fuzzification into several categories must be determined, and is taken at the maximum around 7 in hydrological studies but for preliminary studies it is advised to take at least three fuzzy sets because it can pick the nonlinearity in the variations of the hydrologic phenomenon behavior.
4. The shapes of fuzzy sets in the form of MFs must be determined and it is advised here to depend on triangular and trapezium MFs initially for model establishment. If necessary, later on, the shapes can be changed accordingly. To have a smooth final output product for some problems, one can depend on curvature MFs without corners (continuous), such as sigmoid, Gaussian, etc.
5. In any fuzzification procedure of any linguistic variable, the most right and left side MFs must reach to MDs equal to 1; and if the linguistic variable has open-interval, then end MFs must take the form of trapezium.

The quality of fuzzy set decomposition of linguistic variables according to these steps is very important for successful modeling. Unfortunately, in the open literature, there are many published papers where sometimes the fuzzification stage is not considered properly. It is important to state that at the fuzzification stage, all the fuzzy sets and their MFs are arranged by the researcher, and therefore the fuzzy sets must have normal and convex forms; that is, each fuzzy set should have at least one element with MD equal to 1. For instance, although MF attachment to linguistic variable X as in Figure 6.1 satisfies all the above-mentioned requirements, there is still a conceptual and logical mistake.

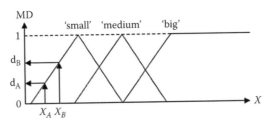

FIGURE 6.1 Input variable MFs.

It is obvious that all the requirements are valid but there is an illogical deduction on the left-hand side MF "small" when X_A and X_B input values are considered. Herein, X_A is smaller than X_B and hence its "small" MD should be greater than X_B, but the opposite is the case as $d_B > d_B$. Such an input fuzzification without careful consideration causes wrong modeling ingredient right from the beginning.

6.3 "IF. . . THEN . . ." RULES

A grammatical sequence of words in any language is a sentence (see Section 2.4 in Chapter 2) but it must imply a meaning by connecting different conditions to each other. The content of a sentence is not grammatically important for a modeler but its content of relationships provides modeling direction. In writing scientific papers, books, or reports, the linguistic grammatical rules are the editors" concern. One should be careful about the scientific content of each sentence. Scientific sentences include two parts, which may be obvious in some cases, but in many others it is not. They are referred to as the statement or premise (see Chapter 3). A premise has two parts: the antecedent (predicate) and consequent sections. In the predicate portion, the input conditions are stated and in the consequent the output results are implied. In the most obvious case, the premise has "IF . . . THEN . . ." structure or some similar construction. The following paragraphs are from the earth sciences literature and in them are scientific sentences that are recognizable by anybody who has logical and rational thinking with the above information in mind.

> The prevailing climate in the southwest of the Kingdom is an interaction of local circulation due to the topography of the area as well as the general circulation which brings moist air from the southern tropical areas close to the equator. Generally, rainfall amount increases "rather sharply" from west to east toward the escarpment and decreases from south to north. However, such variations in the rainfall distribution over the area can be attributed to several factors such as synoptic situation, storm types, seasonal variations, wind speed, orientation of mountains and distance from the moisture sources. (Al-Yamani and Şen, 1997).

> The asymmetry of the probability function of annual sedimentation, whose mode tends towards "low" values but whose variance is such that during one year in 20 on average, the sediment load may be 20 times the median annual value. (AL-Subai, 1991).

> The process of desertification is a "slowly" creeping phenomenon, which takes place in any area during "long" time durations due to different reasons. The first indications are due to variations in the weather or meteorological parameters. In general, desertification implies a decrease in some interested meteorological and agricultural quantities such as the rainfall amounts, vegetation coverage, surface water extensions, groundwater level drops and crop yields. On the other hand, increases in the weather temperature, sand coverage, areal drought coverage, urban area expansion and sedimentation amounts all imply desertification. In simple terms, the historical records of these variables either as time series measurements thorough local land surface instruments or satellite images should lead to increasing or decreasing trends depending on the quantity concerned. (Şen, 2008)

> The flow of water within the Red River is unaffected by the floodway until the discharge reaches 27.432 m³/sec. At this flow the water surface reaches sufficient elevation to permit flow into the floodway. The flow of water into the floodway channel is controlled by a gate structure located downstream of the floodway entrance. The gates, which are

normally flush with the bottom of the river, can be raised to produce a backwater effect, which forces water into the floodway channel. Water is prevented from passing around the control structure and floodway inlet, into the City, by dikes, which extend 43.45 km on either side of the river and floodway entrance. The higher the gates are raised, the greater the backwater and thus the more water are forced into the floodway. By producing a backwater with the gates, the water levels at the upstream communities are increased. The operations policy of the floodway is designed such that under normal conditions the backwater does not alter the upstream water levels in comparison to their natural levels before construction of the floodway. However, in the case of a declared state of emergency, in order to save the downstream City of Winnipeg, flooding of the upstream communities is required (Manitoba Department of Natural Resources, 1984).

"In a serially dependent time series "high" values follow "high" values and "low" values follow "low" values. On the contrary, if "high" values follow "low" values and "low" values follow "high" values, this is a negatively serially dependent series. (Şen, 1974)

"Gavish (1974) revealed that the evaporate minerals are precipitated only in the upper 30–40 cm of the sabkhah (playa), and that from this level down to the ground-water table there is no evaporate accumulation. The salinity of the groundwater in the sabkhah and the concentration of the evaporates in the overlying sediments increase constantly with distance from the shore."

Linguistic means related to language—in our case, plain language words. Examples of linguistic variables are "potential aquifer," "fast speed," "hard rock," "very deep well," "harmless earthquake," "cloudy sky," "weakly active volcano," etc. A fuzzy variable becomes a linguistic variable when it is modified with descriptive words such as *somewhat fast*, *very high*, *really slow*, etc. As already explained in Chapter 2, the main function of linguistic variables is to provide a means of working with the complex systems mentioned above as being too complex to handle by conventional mathematics and formulae. Linguistic variables appear in control systems with feedback loop control and can be related to each other with conditional, "IF . . . THEN . . ." rules. For example,

"IF the water speed is too 'fast' THEN the flood depth is 'big.'"

Any scientific expression with its antecedent and consequent parts implies a relationship between the two, which can be represented as in Figure 6.2 on a Cartesian coordinate system.

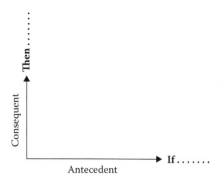

FIGURE 6.2 Predicate relationship.

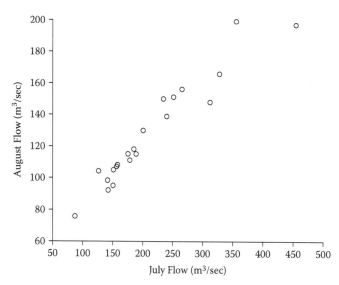

FIGURE 6.3 Successive monthly flows.

Regardless of type, two or more variables are related if in a sample of observations, the values of those variables are distributed in a consistent manner. That is, variables are related if their values correspond systematically to each other for these observations. For example, successive monthly (July versus August) flows at the Ilisu flow measurement station in the southeastern part of Turkey on the Tigris River has a scatter diagram as shown in Figure 6.3, which implies a proportional relationship.

The scatter of points in this figure is consistent and shows a trend that is interpretable in CL as "low" flows in July follow "low" flows in August, and "high" flows in July follow "high" flows in August. The rule base of this statement in the CL context has two rules:

R1: "IF July runoff is 'low' THEN August runoff is 'low.'"

R2: "IF July runoff is 'high' THEN August runoff is 'high.'"

Because the CL is the Law of the Excluded Middle, it is not possible to say anything about the "middle" flows in general, and consequently these two rules are similar to black and white categorization of the monthly flows as directly or indirectly proportional. To indulge gray tones into the relationship, it is necessary to have more than two fuzzy words and with the inclusion of the third "middle" category, as the three fuzzy rules are:

R1: "IF July runoff is 'low' THEN August runoff is 'low.'"

R2: "IF July runoff is 'middle' THEN August runoff is 'low.'"

R3: "IF July runoff is 'high' THEN August runoff is 'high.'"

This last set of fuzzy rules indicates the simplest form of the gray rule base. The grayness of the fuzzy rule base can be increased by increasing the gray fuzzy words in some manner, say, *low middle*, *middle*, and *high middle*, which now yields five rules:

R1: "IF July runoff is 'low' THEN August runoff is 'low.'"

R2: "IF July runoff is 'low middle' THEN August runoff is 'low middle.'"

R3: "IF July runoff is 'middle' THEN August runoff is 'low.'"

R4: "IF July runoff is 'high middle' THEN August runoff is 'high middle.'"

R5: "IF July runoff is 'high' THEN August runoff is 'high.'"

Generally speaking, the ultimate goal of every research or scientific analysis is to find relations between variables. The philosophy of science teaches that there is no other way of representing "meaning" except in terms of relations between some quantities or qualities; either way involves relations rules between variables. Thus, the advancement of science must always involve finding new relations between variables.

The two most elementary formal properties of every relation between variables are the relation's magnitude and its reliability. The magnitude is much easier to understand and to measure than reliability. The reliability of a relation is a much less intuitive concept but still extremely important. It pertains to the "representativeness" of the result found in a specific sample for the entire population. That is, it says how probable is it that a similar relation would be found if the experiment was replicated with other replicates drawn from the same population. Remember that one is almost never "ultimately" interested only in what is going on in his or her replicate (sample) but they are rather interested in the sample only to the extent that it can provide information about the population. If the study meets some specific criteria, then the reliability of a relation between observed variables in the sample can be quantitatively estimated and represented using a standard measurement technique or statistical significance level.

The degree of fuzziness of a system analysis rule can vary from very precise to an opinion held by a human, which would be "fuzzy." Being fuzzy or not fuzzy therefore has to do with the degree of precision of a system analysis rule, which need not be based on human fuzzy perception. For example, the "thicker" the alluvium in a wadi depression, the more the groundwater storage. This can be translated into a fuzzy rule by considering an "IF . . . THEN . . ." statements such as:

"IF alluvium thickness is 'big' THEN the groundwater storage is expected to be 'big.'"

Among many other rules, the following can be counted:

"IF the porosity is 'high' THEN the groundwater storage will be 'high.'"

"IF the rainfall intensity is 'low' THEN the groundwater recharge will be 'small.'"

As the complexity of a system increases, it becomes more difficult and impossible to make a precise statement about its behavior, eventually arriving at a point of complexity where the FL method born in humans is the only way to deal with the problem.

6.4 FUZZY PROPOSITION

Any relationship leads to a conditional statement that if the state of some input variables is such and such, then the output state may be in such a state. In this statement, two words provide the formal way of logical relationships, which are IF and THEN; therefore, from now onward, any relationship in FL terminology will be stated formally in the form of an "IF . . . THEN . . ." sentence. Additionally, in any control or logical statement, human knowledge is also expressed in terms of fuzzy "IF . . . THEN . . ." rules. These are conditional statements such as:

"IF (predicate, antecedent) THEN (consequent)"

In this manner, it is the connection or relationship between the antecedent and consequent parts that implies a relationship between two variables in CL but in fuzzy terminology among the sub-domains (fuzzy sets) of these two or more variables.

The fuzzy statements can be of two types: atomic fuzzy proposition and compound fuzzy proposition. The former is a single statement and the latter is composed of a few single statements that are connected to each other in a serial manner through one of the three logical connectives "AND," "OR," or "NOT." These connectives correspond to union, intersection, and complement operations of classical sets, respectively (see Chapter 3). The fuzzy proposition p_f has two parts separated by a verb in English language. The first part before the verb is the fuzzy variable (subject) f_v, and the second part is the atomic fuzzy word w_f as predicate. Its general structure is:

"f_v is w_f."

There are numerous atomic propositions in water sciences, including:

"Rainfall is 'intensive.'"

"Runoff is 'fast.'"

"Evaporation is 'low.'"

"Humidity is 'high.'"

"Recharge is 'slow.'"

"Hydrograph is 'narrow.'"

Each proposition has an opposite called the negation ("NOTing") of the proposition.
    A compound fuzzy proposition is a composition of atomic (simple) fuzzy proposi-
tions (p_{f1}, p_{f2}, p_{f3}, . . .) that are connected to each other by logical conjunctions. Its
general form can be written as:

"p_{f1} AND p_{f2} OR p_{f3} NOT p_{f4}."

In the following are some of the compound fuzzy hydrologic propositions. Their
number can be expanded to almost an unlimited extent:

"Rainfall is 'slow' AND recharge is 'little.'"

"Evaporation is 'high' AND irradiation is 'extensive' OR humidity is 'medium.'"

"Runoff is 'high' AND demand is 'low' AND storage is 'little.'"

Note that the linguistic variables in a compound statement are not the same.
Furthermore, each atomic proposition in the compound proposition indicates a partial
variation domain of all the linguistic variables considered simultaneously with their
respective fuzzy subsets. Therefore, any compound fuzzy proposition can be considered
a fuzzy relation. The question is how to determine the MFs of these fuzzy relations. The
following answers can be given in light of what has been previously explained.

1. The logical connective "AND" corresponds to intersection, and therefore in
   the calculation of the MD of the compound proposition, the intersection or
   "ANDing" operator starts the execution. A compound proposition such as
   "x is 'A' AND y is 'B'" is interpreted as the fuzzy relation $A \wedge B$ with MFs
   (see Section 3.7.3).
2. If the logical connective is "OR," then the MDs can be computed according
   to the union or "ORing" operator. The general form of "ORing" compound
   proposition is "x is 'A' OR y is 'B'" should be interpreted as the fuzzy relation $A \vee B$ with the corresponding MF calculated from Equation (3.39) in
   Chapter 3:
3. Fuzzy complement operation is used for the logical connective "NOT" in
   a compound proposition. In these propositions, each "NOT" causes the
   replacement of the following set by its complement as explained in Section
   3.7.5 in Chapter 3.

Because n-th fuzzy propositions are interpreted as fuzzy relations, one would like to
know how to interpret "IF . . . THEN . . ." rules. In general, a rule can be written as:

"IF p THEN q."

TABLE 6.1
Truth Values

p	q	"IF p THEN q"
T	T	T
T	F	F
F	T	T
F	F	T

---

where the prepositional variables can be either true, T, or false, F, in classical logic. Table 6.1 shows the final truth value deduced from the truth values of p and q.

Close inspection of Table 6.1 shows that it is possible to express the operation in the classical rule as either

$$\bar{p} \cup q \tag{6.1}$$

or

$$(p \cap q) \cup \bar{p} \tag{6.2}$$

which are equivalent. These two previous equations share the same truth table (Table 6.1).

In a fuzzy rule, p and q propositions are replaced with fuzzy propositions. Hence, the fuzzy rule can be interpreted by replacing the $\overline{}$, $\cup$ and $\cap$ operators in Equations (6.1) and (6.2) with fuzzy complements, union, and intersection, respectively. It was mentioned in Chapter 3 (Section 3.7) that there are a wide variety of fuzzy complements, fuzzy unions (s-norms), and fuzzy intersections (t-norms) operators, a number of different interpretations of the fuzzy rule are proposed in the literature (Ross, 1995). To summarize these operators collectively, p and q are replaced by fuzzy propositions (FP_1) and (FP_2) with the assumption that FP_1 is a fuzzy relation defined in $X = X_1 \times X_2 \times \ldots \times X_n$, and FP_2 in $Y = Y_1 \times Y_2 \times \ldots \times Y_n$. If the fuzzy relation is represented by R, then the Zadeh, Lukasiewicz, Gödel, and Dienes-Rester implications as explained in Chapter 3 are, respectively:

$$\mu_{RZ}(x, y) = \max\{\min[\mu_{FP_1}(x), \mu_{FP_2}(y)], 1 - \mu_{FP_1}(x)\} \tag{6.3}$$

$$\mu_{RL}(x, y) = \min[1, 1 - \mu_{FP_1}(x) + \mu_{FP_2}(y)] \tag{6.4}$$

$$\mu_{RG}(x, y) = \begin{cases} 1 & \text{if } \mu_{FP_1}(x) \le \mu_{FP_2}(y) \\ \mu_{FP_2}(y) & \text{otherwise} \end{cases} \tag{6.5}$$

and

$$\mu_{RD}(x, y) = \max\{1 - \mu_{FP_1}(x), \mu_{FP_2}(y)\} \tag{6.6}$$

On the other hand, most frequently used Mamdani (1974) implications are the minimization of product operators:

$$\mu_{RM}(x, y) = \min[\mu_{FP_1}(x), \mu_{FP_2}(y)] \tag{6.7}$$

or

$$\mu_{RM}(x, y) = \mu_{FP_1}(x)\mu_{FP_2}(y) \tag{6.8}$$

These are the most widely used implications in fuzzy systems and fuzzy control studies (see Chapter 3). Human knowledge in terms of "IF . . . THEN . . ." rules differs from each other, and therefore different implication operations have emerged in practical use. The choice of the most convenient one depends both on the type of problem and on the expert view in problem solving.

On the other hand, some of the terms of "truth" in FL are defined by triangular MFs as in Figure 6.4, which are given an elastic form by hedges (see Section 3.6 in Chapter 3) "fairly true" ("FT"), "fairly false" ("FF"), "very true" ("VT"), and "very false" ("VF"). Unlike the situation in CL, the "truth" values of propositions in FL are allowed to change over the fuzzy sets on the unit interval from 0 to 1, inclusive. For example, a truth value in "FF" appears as a fuzzy set (Figure 6.4).

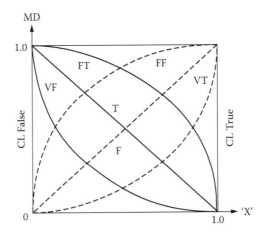

FIGURE 6.4 Fuzzy truth and false.

The truth value of a proposition "f_v is w_f" is a point in a fuzzy set called a linguistic truth value of a fuzzy set "P_1," and is shown symmetrically by $t("P_1")$; then, the following truth values can be written from Chapter 3 as:

$$t(\text{NOT } "P") \rightarrow 1 - t("P") \tag{6.9}$$

$$t("P_1" \text{ AND } "P_2") \rightarrow t("P_1") \wedge t("P_2") = \min\,[t("P_1"), t("P_2")] \tag{6.10}$$

$$t("P_1" \text{ OR } "P_2") \rightarrow t("P_1") \vee t("P_2") = \max\,[t("P_1"), t("P_2")] \tag{6.11}$$

If two propositions "A" and "B" truth values are given as fuzzy sets:

$$t("A") = \left\{ \frac{a_1}{t_{A1}} + \frac{a_{21}}{t_{A2}} + \cdots + \frac{a_n}{t_{An}} \right\}$$

and

$$t("B") = \left\{ \frac{b_1}{t_{B1}} + \frac{b_{21}}{t_{B2}} + \cdots + \frac{b_n}{t_{Bn}} \right\}$$

then consideration of the extension principle (Section 4.8 in Chapter 4) yields:

$$t(\text{NOT } "A") = \left\{ \frac{a_1}{1-t_{A1}} + \frac{a_{21}}{1-t_{A2}} + \cdots + \frac{a_n}{1-t_{An}} \right\} \tag{6.12}$$

$$t("A" \text{ AND } "B") = t("A") \wedge t("B") = \sum_{i,j}^{n} \frac{\min(a_i, b_j)}{\min(t_{Ai}, t_{Bj})} \tag{6.13}$$

$$t("A" \text{ OR } "B") = t("A") \vee t("B") = \sum_{i,j}^{n} \frac{\min(a_i, b_j)}{\max(t_{Ai}, t_{Bj})} \tag{6.14}$$

Example 6.1

Let the truth values of two propositions "A" and "B" be given as:

$$t("A") = t("\text{almost true}") = \left\{ \frac{0.6}{0.8} + \frac{1.0}{0.9} + \frac{0.5}{1.0} \right\}$$

$$t("B") = t("\text{more or less true}") = \left\{ \frac{0.5}{0.6} + \frac{0.7}{0.7} + \frac{1.0}{0.8} + \frac{1.0}{0.9} + \frac{1.0}{1.0} \right\}$$

TABLE 6.2

Fuzzy Truth States

Truth Value	"ANDing"			"ORing"		
	T	F	T + F	T	F	T + F
T	T	F	T + F	T	T	T
F	F	F	F	T	F	T + F
T + F	T + F	F	T + F	t	T + F	T + F

Hence,

$$t(\text{"}B\text{" AND "}B\text{"}) = t(\text{"almost true" AND "more or less true"}) = \left\{ \frac{0.5}{0.6} + \frac{0.7}{0.7} + \frac{1.0}{0.8} + \frac{1.0}{0.9} + \frac{1.0}{1.0} \right\}$$

$$t(\text{"}B\text{" OR "}B\text{"}) = t(\text{"almost true" OR "more or less true"}) = \left\{ \frac{0.6}{0.8} + \frac{1.0}{0.9} + \frac{1.0}{1.0} \right\}$$

The Zadeh (1999) truth table for "ANDing" (conjunction) and "ORing" (disjunction) are given in Table 6.2, where T + F implies an "unknown" state in a fuzzy manner. For approximate reasoning, FL allows the use of four different types—fuzzy predicates (FPRs), fuzzy modifiers (FMOs), fuzzy quantifiers (FQUs), and fuzzy qualifiers (FQAs)—in any proposition. These help to distinguish between CL and FL:

1. FPRs: This corresponds to the case where the predicates can be fuzzy sets. Hence, there may be a fuzzy proposition such as "The river is 'long.'" It is worth noting that most of the predicates in natural language are fuzzy rather than crisp.

2. FMOs: These are fuzzy predicate modifiers and in CL the only widely used modifier is the negation "NOT." In FL, in addition to this there are many modifiers as explained in Chapter 3 (Section 3.6) under the title of hedges. For instance, the proposition as "The rainfall was 'extremely intensive' has a modifier of 'extremely.'"

3. FQUs: Among the most widely used fuzzy quantifiers are "many," "several," "most," "few," "frequently," "about," "occasionally," and "most." In FL context, an FQU is like a fuzzy number or a fuzzy proposition that provides imprecision in one or more fuzzy or crisp sets. For instance, ""Most extensive" drainage basins are "dry"" can be interpreted as fuzzy proposition of the fuzzy set of "dry drainage basins" in the fuzzy set of "most extensive." Additionally, FQUs can also be used to represent the propositions" meaning implying probabilities and hence make it possible to manipulate probabilities with FL. For example, "Most floods are 'very dangerous'" implies conditional probability in the sense that among the observed floods (perhaps a certain number of events), a certain proportion (hence percentage) is dangerous. Hence, fuzzy probabilities and FQUs are interchangeable and therefore any proposition involving fuzzy probabilities may be replaced by a semantically equivalent proposition involving FQUs, which means that identical probability distributions can be induced.

4. FQAs: In general, FL has four major qualification modes as follows:
   a. *Fuzzy truth qualification:* The propositions expressed as "(P) is (t)" has a truth value (t). For example, "(Tigris river is 'short') is (not 'very true')" has the qualified proposition of (Tigris river is "short") and the qualifying fuzzy truth value of (not "very true"),
   b. *Fuzzy probability qualification:* This has the general form of "(P) is p," where p is a fuzzy probability. It is well known that in the CL, probability is a crisp number or interval but that in FL, there are different alternatives for fuzzy probabilities depending on qualifiers such as "likely," "probably," "unlikely," "about," etc. For instance, "(Tigris river is 'long') is ('very likely')." Such fuzzy probabilities may be considered fuzzy numbers and hence fuzzy arithmetic can be applied as in Chapter 4.
   c. *Fuzzy possibility qualification:* Among such qualifiers are "possible," "impossible," and various hedges attached to these words. For example, "(Tigris river is 'short') is ('almost impossible')" includes "almost impossible" as a qualifying fuzzy possibility.
   d. *Fuzzy usuality quantification:* The general form of such a qualifier is "usually ('X' is 'F')," where the variable "X" is a subject in a universe of discourse U and the predicate F is a fuzzy subset of the same universe, which may be interpreted as a usual value of "X" denoted as U("X") = "F." The concept of usuality relates to propositions that are usually true or to events that have a high probability of occurrence.

Example 6.2

If s and i are the groundwater speed and hydraulic gradient, respectively, and q is the specific discharge linguistic variable, then the following fuzzy "IF . . . THEN . . ." relationship is valid.

"IF s is 'slow' and i is 'small' THEN q is 'low.'"

where "slow" and "small" are the fuzzy subsets defined in Figure 6.5 as a trapezium and triangle individually, but they specify a certain sub-area on the variation diagram.

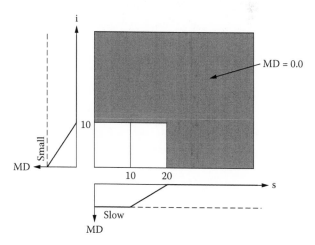

FIGURE 6.5 Illustration of $\mu_{slow}(s)$ and $\mu_{small}(i)$.

Accordingly, the fuzzy expression of "slow" is given as:

$$\mu_{slow}(s) = \begin{cases} 1 & \text{if } s \le 10 \\ \dfrac{20-s}{10} & \text{if } 10 < s \le 20 \\ 0 & \text{if } s > 20 \end{cases} \tag{6.15}$$

Additionally, "small" is the fuzzy set in the domain of hydraulic gradient with the following MF:

$$\mu_{small}(i) = \begin{cases} \dfrac{10-i}{10} & \text{if } 0 \le i \le 10 \\ 0 & \text{if } i > 10 \end{cases} \tag{6.16}$$

Finally, "low" is in the specific discharge domain with the following MF:

$$\mu_{low}(q) = \begin{cases} 0 & \text{if } q \le 1 \\ q-1 & \text{if } 1 < q \le 2 \\ 1 & \text{if } q > 2 \end{cases} \tag{6.17}$$

The variation domains of s, i, and k are [0, 20], [0, 10], and [0, 2], respectively. According to the algebraic product for t-norm, the fuzzy proposition becomes $FP_1 = $ "s is 'slow' and i is 'small,'" which represents the fuzzy relation in the universe of discourses as shown in Figure 6.5 and with the following MF:

$$\mu_{FP_1}(s,i) = \mu_{slow}(s)\mu_{small}(i)$$

$$= \begin{cases} 0 & \text{if } s \ge 20 \text{ or } i \ge 10 \\ \dfrac{10-i}{10} & \text{if } s \le 10 \text{ and } i \le 10 \\ \dfrac{(20-s)(10-i)}{200} & \text{if } 20 < s \le 20 \text{ and } i \le 10 \end{cases} \tag{6.18}$$

Example 6.3

Sometimes it is possible to present the fuzzy subset MFs by convenient mathematical expressions such as (Wang and Mendel, 1992a):

$$\mu_{slow}(s) = \dfrac{1}{1+e^{\frac{s-45}{5}}} \tag{6.19}$$

$$\mu_{small}(i) = \dfrac{1}{1+e^{\frac{i-5}{2}}} \tag{6.20}$$

and

$$\mu_{low}(q) = \frac{1}{1+e^{2(-q+1.25)}} \tag{6.21}$$

If the Mamdani product implication and algebraic product for the *t*-norms are used, then the composition membership can be found as:

$$\mu_{QMP}(s,i,k) = \mu_{slow}(s)\mu_{small}(i)\mu_{low}(q)$$

$$= \frac{1}{\left(1+e^{\frac{s-45}{5}}\right)\left(1+e^{\frac{i-5}{2}}\right)\left(\frac{1}{1+e^{2(-q+1.25)}}\right)} \tag{6.22}$$

Example 6.4

If the rainfall *R* and evaporation *E*, universal sets are $R = \{1, 2, 3, 4\}$ and $E = \{1, 2, 3\}$, respectively, and the fuzzy rule is:

"IF rainfall is 'large' THEN evaporation is 'small.'"

with two atomic fuzzy words:

"large" = {0/1 + 0.1/2 + 0.5/3 + 1.0/4}

and

"small" = {1/1 + 0.5/2 + 0.1/3}

then find the implications according to Dienes-Rescher (I_D), Lukasiewicz (I_L), Zadeh (I_Z), Gödel (I_G), and Mamdani maximum (I_{MM}) and product (I_{MP}). These implications are, respectively:

I_D = 1/(1,1) + 1/(1, 2) + 1/(1, 3) + 1/(2, 1) + 0.9/(2, 2) + 0.9/(2, 3) + 1/(3, 1)

   + 0.5/(3, 2) + 0.5/(3, 3) + 1/(4, 1) + 0.5/(4, 2) + 0.1/(4, 1)

I_L = 1/(1, 1) + 1/(1, 2) + 1/(1, 3) + 1/(2, 1) + 1/(2, 2) + 1/(2, 3) + 1/(3, 1)

   + 1/(3, 2) + 0.6/(3, 3) + 1/(4, 1) + 0.5/(4, 2) + 0.1/(4, 3)

I_Z = 1/(1, 1) + 1/(1, 2) + 1/(1, 3) + 0.9/(2, 1) + 0.9/(2, 2) + 0.9/(2, 3) + 0.5/(3, 1)

   + 0.5/(3, 2) + 0.5/(3, 3) + 1/(4, 1) + 0.5/(4, 2) + 0.1/(4, 3)

I_G = 1/(1, 1) + 1/(2, 1) + 1/(1, 3) + 1/(2, 2) + 1/(2, 3) + 1/(3, 1) + 1/(3, 2)

   + 0.1/(3, 3) + 1/(4,1) + 0.5/(4, 2) + 0.1/ (4, 3)

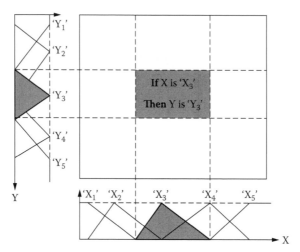

FIGURE 6.6 One rule base defined a sub-area from the variation domain.

and

$$I_{MM} = 0/(1, 1) + 0/(1, 2) + 0/(1, 3) + 0.1/(2, 1) + 0.1/(2, 2) + 0.1/(2, 3)$$

$$+ 0.5/(3, 1) + 0.5/(3, 2) + 0.1/(3, 3) + 1/(4, 1) + 0.5/(4, 2) + 0.1/(4, 3)$$

$$I_{MP} = 0/(1, 1) + 0/(1, 2) + 0/(1, 3) + 0.1/(2, 1) + 0.05/(2, 2) + 0.01/(2, 3)$$

$$+ 0.5/(3, 1) + 0.25/(3, 2) + 0.05/(3, 3) + 1/(4, 1) + 0.5/(4, 2) + 0.1/(4, 3)$$

Mamdani implications are local whereas the others are all global. After all that has been explained in this and previous sections, it is possible to visualize that any single rule defines a definite sub-area on the variation domain as shown in Figure 6.6.

However, a collection of different rules defines a joint partial variation domain as presented in Figure 6.7 which is similar to that explained in Chapter 5 (Section 5.6.1).

6.5 INPUT RULE BASE ESTABLISHMENT

Rules are a set of "IF . . . THEN . . ." statements that combines logical fuzzy sets of input variables in a meaningful manner if the underlying relationship is perceived by the researcher, planner, or modeler hydrologist. First, a set of rules represents the internal associations between input variables, and then such combinations are related to output fuzzy sets through logical proportional, indirectly proportional, or other type of relationship (as discussed in Chapter 5). Prior to the inference for the estimation or estimation of the output variable, under a certain set of input data, the rule base must be constructed by all means. Among these means there are four possible techniques for increasing the reliability of hydrologic event representation: mechanical, personal intuition, expert view, and data elimination. Finally, the representative (valid) set of rules is obtained for the hydrologic phenomenon

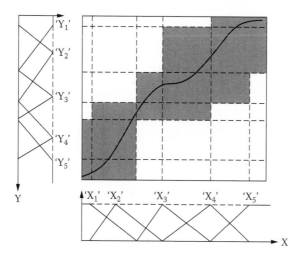

FIGURE 6.7 All rule bases defined as sub-areas from the variation domain.

under consideration. The overall decision is based on a *set* of rules such that *all* the applicable rules are invoked using input MFs and, if available, data in order to determine the validity of each rule, and they are mapped in turn into MFs and truth-value controlling the output variable. The fuzzy rule base includes all the features of mathematical formulation of an event but in terms of words and propositions in the form of "IF . . . THEN . . ." statements. This is very similar to an algorithm development procedure in computer software. Without knowing the philosophy and logic behind the event generation mechanism, it is not possible to write computer software; and likewise, without writing down all the possible FL relationships (rules) between the input variables and their combination with the output variable fuzzy sets, it is not possible to solve any problem by FL algorithm or modeling. A fuzzy algorithm, then, is a procedure, usually a computer software program, made up of rules relating linguistic variables. For instance, "IF rainfall is 'very intensive' THEN evaporation is 'much smaller'" and "IF the rate of change in temperature of the air is 'much too high' THEN turn the heater 'down a lot'" are simple rules.

After the fuzzification procedure, the input and output variables are identified linguistically in terms of fuzzy sets for human and MFs for computer and now it is necessary to first combine the input variables among themselves and then the output variable fuzzy sets can be attached logically to the composite form of input sets. In the rule base, establishing the following points is noteworthy:

1. Logical combination of the input fuzzy sets are combined together with the "ANDing" operation which implies that in each rule, only one fuzzy set from each linguistic input variable must be considered. If there are m linguistic input variables, then each rule will include $(m-1)$ "ANDing" operations in the antecedent part. Hence, the antecedent part of the rule is filled with propositions.

2. After the whole combination of possible alternatives for the problem at hand, there appears a set of rules in the form of "IF . . . THEN . . ." statements without any specification of the consequent part of the rule.
3. Each rule is attached to other rules in the rule base with an "ORing" operation; and hence, if there are k rules, then there will be $(k − 1)$ "ORing" operators.
4. The input variables are the conditions (reasons) for the generating mechanism in the problem and, hence, their logical identification is necessary prior to any consequent part attachment.

There are different ways of identifying the rule base of a phenomenon and in sequence of information content increment, they are mechanical documentation, personal intuition, expert view, and database searches.

6.5.1 Mechanical Documentation

If the fuzzy modeler does not know any information about the phenomenon, then he/she can write the rule base without exception in a mechanical manner. For this purpose, it is necessary to know the names of the linguistic input variables and their fuzzy set granulations (i.e., fuzzy set names and numbers). If there are m input variables $(I_1, I_2, \ldots, I_m)$ each with $n_1, n_2, \ldots, n_m$ fuzzy sets as granulation (see Chapter 5), then the exhaustive and mutually inclusive combinations of all these fuzzy sets will be $k = n_1 \times n_2 \times \ldots \times n_m$ rules within the rule base. Such an approach divides the m dimensional input variation domain of the variables into k sub-domains (categories) with overlappings $(I_{1i}, I_{2j}, \ldots, I_{ml})$. This is tantamount to saying that none of these sub-domains are preferable over the others because there is as yet no logical or expert information injected into the establishment of the rule base. It is known that if the input variables and their sub-domains (fuzzy sets) are not in a harmonious relationship, then this rule will not be considered a member in the rule base. There is no role for the output variable in the identification of the valid rules at this stage. Hence, in mechanical rule base writing the general form of the base will be as follows:

R1: "IF I_{11} is '. . .' AND I_{21} is '. . .' AND . . . AND I_{ml} is '. . .' THEN . . ."

OR

Ri: "IF I_{1i} is '. . .' AND I_{2j} is '. . .' AND . . . AND I_{ml} is '. . .' THEN . . ."

OR

Rk: "IF I_{1n1} is '. . .' AND I_{2n2} is '. . .' AND . . . AND I_{mnm} is '. . .' THEN . . ."

Because there is no preference of any input sub-domain over the others, the modeler is completely ignorant and this means that the occurrence of the phenomenon is treated as completely independent so that in each sub-domain there is some input fuzzy set combination. Of course, this is not quite possible in hydrological events, which have interrelations to a certain extent and this point indicates that some of the rules from mechanical documentation must be eliminated. In the process of elimination of rules from the rule base, there are three processes in sequence:

1. Human experts are generally able to provide certain linguistic information. This is even true for formally uneducated people like the shepherd in Chapter 2. However, human experts are usually unable to provide a full account of their practical knowledge and experience in terms of linguistic rules alone. Consequently, some information is lost.
2. Additional information can be gained through instrumental measurements in either the field or laboratory.
3. Some statistical procedures provide treated information. Although the input-output data pairs provide useful information for the generation or elimination of rules, they may not be sufficient in all the cases.

These processes may not cover a large variety of situations, and therefore some of the rules may not be valid. In fuzzy rule base generation, both linguistic and numerical data are of interest. Wang and Mendel (1992b) suggested such a method and it is repeated here as an example in hydrology.

Example 6.5

The groundwater recharge G_R is a function of rainfall R and infiltration I. Herein, groundwater recharge is the consequent variable whereas rainfall and infiltration are input variables. Hence, the problem is a two-input-one-output modeling with rainfall and infiltration antecedent variables. If the rainfall is fuzzified into three fuzzy sets as "light," "middle," and "heavy" and similarly infiltration into "low," "medium," and "high" fuzzy sets, then there are $3^2 = 9$ rules without any distinction and it is therefore possible to write the rule base automatically as:

R1: "IF rainfall is 'light' AND infiltration is 'low' THEN . . ."

R2: "IF rainfall is 'light' AND infiltration is 'medium' THEN . . ."

R3: "IF rainfall is 'light' AND infiltration is 'high' THEN . . ."

R4: "IF rainfall is 'middle' AND infiltration is 'low' THEN . . ."

R5: "IF rainfall is 'middle' AND infiltration is 'medium' THEN . . ."

R6: "IF rainfall is 'middle' AND infiltration is 'high' THEN . . ."

R7: "IF rainfall is 'heavy' AND infiltration is 'low' THEN . . ."

R8: "IF rainfall is 'heavy' AND infiltration is 'medium' THEN . . ."

R9: "IF rainfall is 'heavy' AND infiltration is 'high' THEN . . ."

6.5.2 PERSONAL INTUITION

Herein, personal logic does not mean that the person is outside the speciality completely but he/she may not be expert enough or there may be experts better than him/her in the phenomenon of concern. The fuzzy logic modeler must depend on his/her plain logical facts and also on the speciality information. Such information sources and experience may help the modeler delete some of the rules from the antecedent parts only. For instance, if among the input linguistic variables are "temperature" (T), "rainfall" (R), and "humidity" (h), with their fuzzy granulations as "low," "middle," and "high" then in this case it is possible to cancel any rule with "IF T is 'high' AND R is 'high' THEN . . ." because "high" temperature does not lead to "high" precipitation. Likewise, think about the rule "IF T is "high" AND h is "low" THEN . . .," etc.

This point indicates that the hydrologist can eliminate many illogical rules just from the antecedent part by thinking about the rational and logically possible combinations of inputs pairwise only. Of course, this can also be done for three inputs simultaneously, if possible. In this manner, personal logic and experience will reduce the number of rules in the mechanical rule base documentation.

Example 6.6

Try to eliminate some of the mechanical rules in Example 6.5 by personal logical thinking by considering the possibility of rainfall and infiltration relationship. For instance, logically, it is possible to expect neither ""high" infiltration after a "light" rainfall" nor ""low" infiltration after a "heavy" rainfall," depending on the antecedent field conditions. Hence, these two illogical statements imply that R3 and R7 cannot be valid in this problem of groundwater recharge.

6.5.3 EXPERT VIEW

The reason why FL principles cited in the literature are counted among the expert systems is due to the fact that the rule base is established with expert views also. Every expert will have his/her own way of logical and expertise judgment, but even experts in the same speciality might differ in some rules, whereas many of the rules remain as they are and in some of the rules within the rule base there might be less emphasis. This point indicates that each expert might give different weight of each rule and hence within the same rule base every rule may not have the same weight. This is expressed as the "degree of belief" in the literature (Ross, 1995). It is interesting to note that even the belief of the expert enters into the FL modeling procedure. However, in practice, most often the degree of belief is considered as 1 for each rule, and therefore they are equally treated in the overall modeling.

After the mechanical and personal intuition rule elimination stages, the remaining rules are subjected to the review of one or more experts and finally the consensus is satisfied by all parties and this will be the rule base that is based on knowledge, information, experience, logic, and expertise. So far, all the rule base development stages are without consideration for the database. Hence, if one goes through all these stages, one becomes linguistically expert about the generating mechanism of the phenomenon without data but by knowing the logical composition of the problem instead of the mathematical equations, which may lead one to memorization. The hydrologist should skip mechanical thinking and memorization, which refrain the rule base development abilities.

Example 6.7

The rules in the rule base after Example 6.6 can be eliminated by taking into consideration the expert view of the hydrologist. For instance, if in the study area, previous assessments have indicated that even in the cases of "light" rainfall there is no infiltration and also "middle" rainfall does not lead to "medium" infiltration, then rules R1 and R5 should be discarded.

6.5.4 DATABASE SEARCH

It is possible to say that data is the expert of experts provided there are no mistakes in that data. A database is needed right from the beginning, especially in probabilistic, statistical, stochastic, and mathematical modeling, to be used for readily available formulations without much logical basis or with CL foundation. However, in FL modeling, the data is necessary during or after the rule base configuration in light of the aforementioned stages. Then this obtained rule base will be subjected to available numerical data through fuzzy sets (MFs), and the range of variability of linguistic variables must also be expressed in numbers. Such numbering will provide a common base for comparison, and the rules in the rule base will be tested with the data for their validity. For any rule to be valid, again its antecedent part with "ANDing"s is compared with the data. If the rule is valid, this means that the data is triggering (firing) the antecedent part of the rule without exception of any linguistic variable fuzzy set.

6.5.4.1 Triggering

To understand the triggering procedure, it is necessary to consider rules in their geometrical shapes for humans to understand and in mathematical formulations (MFs) for computers to make the calculations. Suppose there are two input variables "X" and "Y" with three fuzzy sets each. In this case, there will be $3 \times 3 = 9$ mechanical rules. Provided the data is at hand, it may not be necessary to go through the personal intuition and expert view stages because the data will identify objectively the whole rule base. However, the hydrologist would do well not to ignore these two stages if he/she is willing to train himself/herself in the proper direction. The mechanical rule base ingredients with all MFs are given in Figure 6.8.

Figure 6.9 indicates all the mechanical rules collectively in the form of explicit MFs that have been combined in an exhaustive and mutually exclusive manner.

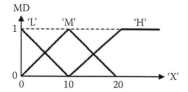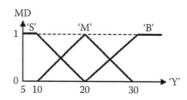

FIGURE 6.8 Two input MFs.

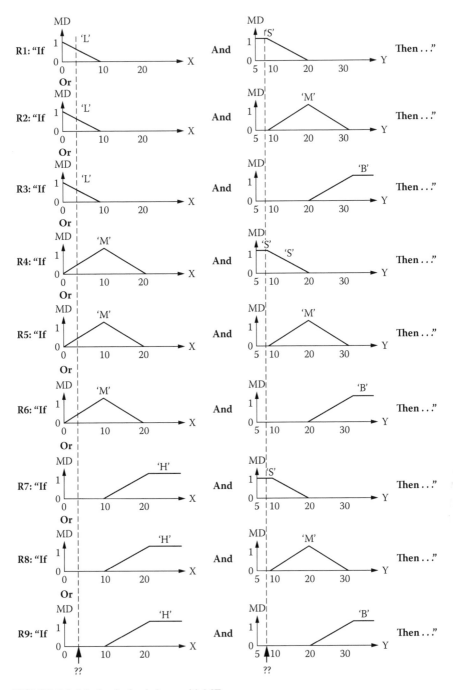

FIGURE 6.9 Mechanical rule base with MFs.

TABLE 6.3
Input and Output Data

Data Number	"X"	"Y"	"Z"	Trigger	Output MFs
1	3.5	7.8	16.5	R1, R4	"LM"
2	6.7	12.5	22.3	R1, R2, R4, R5	
3	1.5	3.4	11.5	R1, R4	
4	21.2	15.1	19.3	R8	
5	17.0	27.3	32.7	R5, R6, R8, R9	

Because many rules in the variation domain of "X" and "Y" are triggered, they do not indicate a consistent trend and consequently the data is expected to be scattered randomly.

In hydrological studies, the data is given as in Table 6.3; and although the number of data may be a lot, herein only five hypothetical data values are considered. In the same table, the output data "Z" is also given.

After the entrance of the first data point into Figure 6.9, it is obvious that the first data fires rules that are shown in the trigger column of Table 6.3. To trigger a rule, both of the MFs from "X" and "Y" must be fired simultaneously. Consideration of this point indicates that for the first data value, R1 and R4 are triggered. It is possible to complete the table with other input data values in the same manner. In doing so according to the data at hand, one is able to identify all the valid rules from the rule base and some of the rules will not be triggered by any data and therefore must be eliminated, and finally, relevant rules will remain in the rule base. With the given data at hand, R3 is not triggered at all and hence it must be eliminated from the rule base.

6.5.4.2 Degree of Belief

After the completion of Table 6.3 with triggered rules, it is possible to calculate the percentage of each fired rule out of the total triggers. Of course, the ones that are not triggered at all have a zero degree of belief for the problem at hand, and they are eliminated already. The total number of triggered rules N_R and the number of each triggered rule N_{Ri} ($i = 1, 2, \ldots, k$) can be counted from the trigger column in Table 6.3. Finally, the degree of belief B_i or fire (trigger) percentage of each rule can be calculated as:

$$B_i = \frac{N_{Ri}}{N_R} \tag{6.23}$$

The summation of the degrees of belief is equal to 1. In a way, the degree of belief indicates the histogram of rule base triggering percentages. For instance, in Table 6.3 there are 11 triggerings, R4 is triggered twice, and hence its degree of belief is $2/11 = 0.18$. Hence, one can identify that each rule has a different degree of belief and none of them can have a degree of belief equal to 1, which is the case in CL deductions where, for example, one can say:

"IF rainfall increases THEN runoff increases."

which does not include any fuzziness in its content but crispness prevails as the rule for the whole rainfall-runoff phenomenon. It is possible to deduce that in CL studies such as formulations (equations), there is only one result and it is to be believed that it is true, which is not acceptable in the context of fuzzy modeling.

6.6 COMPLETE RULE BASE

After the identification of the rule base components for a given event, it is time to attach the output fuzzy sets (MFs) for the completion of the rule base prior to make output estimations. It is not possible to attach the output MFs mechanically because there is no logical rule base for the event generation mechanism. It is possible that personal intuition and experience may fill partially the last part (consequent) in each rule; other partial rules may be filled by expert views; and finally, in the case of database availability, it is now possible to fill the last part of each rule objectively by the use of data. The first step in such a procedure is to granulate the output linguistic variable into a set of fuzzy sets (MFs). Because in practical studies refined estimations or predictions are required, it is usual to divide the output linguistic variable into more than input variable fuzzy sets. Practical advice suggests not to increase the number of fuzzy sets in the output linguistic variable by more than seven. However, the more the data, the more can be the number of output fuzzy sets. Figure 6.10 shows such a granulation of the linguistic output variable into five MFs. In this figure, "L," "LM," "M," "HM," and "H" mean "low," "low medium," "medium," "high medium," and "high" fuzzy sets or MFs, respectively.

By considering the available data, one can fill the last column in Table 6.3 with output MFs. For instance, for the first data set, $Z_1 = 16.5$, the entrance of the Z axis leads to the coincidence with "M" and "LM" MFs, which have MDs as d_l and d_u, respectively. Because there are two MDs, it is necessary to make the "ORing" operation and hence the one with the bigger MD is considered the right-hand side of the rule. That is to say, the valid MF for the last part of the rule is "LM." Hence, rules R1 and R4 will have their complete forms as follows:

R1: "IF X is 'low' AND Y is 'small' THEN Z is 'low medium'"

or

R4: "IF X is 'medium' AND Y is 'small' THEN Z is 'low medium'"

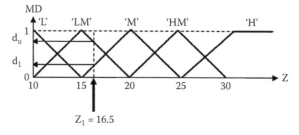

FIGURE 6.10 Output MFs.

TABLE 6.4
Data Set

Data Number	"X_1"	"X_2"	"Y"	Data Number	"X_1"	"X_2"	"Y"
1	140	24	19	14	−60	−34	61
3	−140	−24	81	15	50	0	39
3	130	41	19	16	−50	0	61
4	−130	−41	81	17	45	−22	33
5	120	43	25	18	−45	22	67
6	−120	−43	75	19	30	10	43
7	83	0	31	20	−30	−10	57
8	−83	0	69	21	95	70	34
9	90	−1	40	22	−95	−70	66
10	−90	1	60	23	80	77	36
11	150	83	22	34	−80	−77	64
12	−150	−83	78	25	90	14	39
13	60	34	39	26	−90	−14	61

After the completion of the rule base including output MFs, the hydrologist is now ready to make an estimation or a prediction of the output value from the rule base through some fuzzy inference systems (FISs), as explained in Chapter 7.

Example 6.8

Let the two input variables be "X_1" and "X_2" and the corresponding output is denoted by "Y." The simultaneous measurements of these hydrological variables led to 26 data sets as in Table 6.4. Each data set is shown by (X_{1i}, X_{2i}, Y_i) where $i = 1, 2, \ldots, 26$.

The following steps are necessary for the construction of the relevant rule base for the problem at hand.

1. The first decision concerns the determination of each input and output variability domain. It is shown in Table 6.5.
2. Divide each interval into m sub-classes where m can be different for different variables. For instance, in Figure 6.11 the input variable "X_1" is divided into nine classes.

TABLE 6.5
Variable Variation Domain

Variable	Domain Interval
"X_1"	[−200, 200]
"X_2"	[−100, 100]
"Y"	[0, 100]

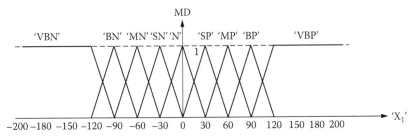

FIGURE 6.11 Input variable division.

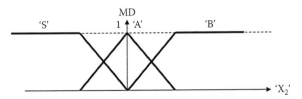

FIGURE 6.12 Input variable division.

In Figure 6.11, "VBN," "BN," "MN," "SN," "N," "SP," "MP," "BP," and "VBP" fuzzy sets imply "very big negative," "big negative," "middle negative," "small negative," "negative," "small positive," "middle positive," "big positive," and "very big positive" MFs, respectively. Likewise, the second input variable fuzzi-fication is represented by three words: *small* ("S"), *average* ("A"), and *big* ("B") (Figure 6.12).

Finally, the output linguistic variable "Y" has nine MFs: "very very small" ("VVS"), "very small" ("VS"), "medium small" ("MS"), "small" ("S"), "medium" ("M"), "big" ("B"), "medium big" ("MB"), "very big" ("VB"), and "very very big" ("VVB") (Figure 6.13). Mechanically, there are 27 rules.

The valid triggering rules, given the data in Table 6.4 and MFs in Figures 6.11 through 6.13, are presented in Table 6.6.

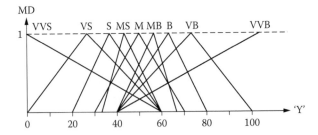

FIGURE 6.13 Output variable MFs.

TABLE 6.6
Rules from Input-Output Data

Rule Number	Input-Output Data Pair	Rule
R1	$X_{1,1}, X_{2,1}, Y_1$	IF X_1 is "VBP" and X_2 is "ASF" THEN Y is "VVS."
R2	$X_{1,2}, X_{2,2}, Y_2$	IF X_1 is "VBN" and X_2 is "ASF" THEN Y is "VVB."
R3	$X_{1,3}, X_{2,3}, Y_3$	IF X_1 is "VBP" and X_2 is "BF" THEN Y is "VVS."
R4	$X_{1,4}, X_{2,4}, Y_4$	IF X_1 is "VBN" and X_2 is "SF" THEN Y is "VVB."
R5	$X_{1,5}, X_{2,5}, Y_5$	IF X_1 is "VBP" and X_2 is "BF" THEN Y is "VS."
R6	$X_{1,6}, X_{2,6}, Y_6$	IF X_1 is "VBN" and X_2 is "SF" THEN Y is "VVB."
R7	$X_{1,7}, X_{2,7}, Y_7$	IF X_1 is "BP" and X_2 is "ASF" THEN Y is "VS."
R8	$X_{1,8}, X_{2,8}, Y_8$	IF X_1 is "BN" and X_2 is "ASF" THEN Y is "VB."
R9	$X_{1,9}, X_{2,9}, Y_9$	IF X_1 is "BP" and X_2 is "ASF" THEN Y is "S."
R10	$X_{1,10}, X_{2,10}, Y_{10}$	IF X_1 is "BN" and X_2 is "ASF" THEN Y is "B."
R11	$X_{1,11}, X_{2,11}, Y_{11}$	IF X_1 is "VBP" and X_2 is "BF" THEN Y is "VS."
R12	$X_{1,12}, X_{2,12}, Y_{12}$	IF X_1 is "VBN" and X_2 is "SF" THEN Y is "VB."
R13	$X_{1,13}, X_{2,13}, Y_{13}$	IF X_1 is "MP" and X_2 is "BF" THEN Y is "S."
R14	$X_{1,14}, X_{2,14}, Y_{14}$	IF X_1 is "MN" and X_2 is "SF" THEN Y is "B."
R15	$X_{1,15}, X_{2,15}, Y_{15}$	IF X_1 is "MP" and X_2 is "ASF" THEN Y is "S."
R16	$X_{1,16}, X_{2,16}, Y_{16}$	IF X_1 is "MN" and X_2 is "ASF" THEN Y is "B."
R17	$X_{1,17}, X_{2,17}, Y_{17}$	IF X_1 is "MP" and X_2 is "ASF" THEN Y is "VS."
R18	$X_{1,18}, X_{2,18}, Y_{18}$	IF X_1 is "MN" and X_2 is "ASF" THEN Y is "VB."
R19	$X_{1,19}, X_{2,19}, Y_{19}$	IF X_1 is "SP" and X_2 is "ASF" THEN Y is "MS."
R20	$X_{1,20}, X_{2,20}, Y_{20}$	IF X_1 is "SN" and X_2 is "ASF" THEN Y is "MB."
R21	$X_{1,21}, X_{2,21}, Y_{21}$	IF X_1 is "BP" and X_2 is "BF" THEN Y is "VS."
R22	$X_{1,22}, X_{2,22}, Y_{22}$	IF X_1 is "BN" and X_2 is "SF" THEN Y is "VB."
R23	$X_{1,23}, X_{2,23}, Y_{23}$	IF X_1 is "BP" and X_2 is "BF" THEN Y is "S."
R24	$X_{1,24}, X_{2,24}, Y_{24}$	IF X_1 is "BN" and X_2 is "SF" THEN Y is "B."
R25	$X_{1,25}, X_{2,25}, Y_{25}$	IF X_1 is "BP" and X_2 is "ASF" THEN Y is "S."
R26	$X_{1,26}, X_{2,26}, Y_{26}$	IF X_1 is "BN" and X_2 is "ASF" THEN Y is "B."

REFERENCES

Al-Suba'i, K.A.M.G., (1991). Erosion-sedimentation and seismic considerations for dam siting in the central Tihamat Asir region. Unpublished Ph.D. Thesis, King Abdulaziz University, Faculty of Earth Sciences Kingdom of Saudi Arabia, 343 pp.

Al-Yamani, M.S. and Şen, Z. 1997. Spatiotemporal dry and wet spell duration distributions in southwestern Saudi Arabia. *Theoret. Appl. Climatol.* 57(3–4): 165–179.

Gavish, E. 1974. Geochemistry and mineralogy of a recent sabkhah along the coast of Sinai, Gulf of Suea, *Sedimentology* 21: 397–414.

Mamdani, E.H. 1974. Application of fuzzy algorithms for simple dynamic plant. *Proc. IEE* 121: 1585–1588.

Manitoba. Dept. of Natural Resources, 1984. Natural Resources and Environment. Annual report, Dept. of Natural Resources, Box 22 Winnipeg, Man. R3H OW9.

Ross, J.T. 1995. *Fuzzy Logic with Engineering Applications.* McGraw-Hill, New York, 593 pp.

Şen, Z., 1974. Small sample properties of stationary stochastic models and the Hurst phenomenon in Hydrology. Unpublished Ph.D. Thesis, Imperial College of Science and Technology, University of London, 286 pp.

Şen, Z., 2008. Wadi Hydrology. Taylor & Francis Group, CRC Press, Boca Raton, 347 pp.

Wang, L.X. and Mendel, J.M. 1992a. Fuzzy basis functions, universal approximation, and orthogonal least squares learning. *IEEE Trans. on Neural Network* 3(5): 807–814.

Wang, L.X. and Mendel, J.M. 1992b. Generating fuzzy rules by learning from examples. *IEEE Trans. Systems, Man, and Cybernet.* 22(5): 1414–1427.

Zadeh, L.A. 1999. From computing with numbers to computing with words—From manipulation of measurements to manipulation of perceptions. *IEEE Trans. Circuits and Systems* 45(1): 105–119.

PROBLEMS

6.1 Provide three assumptions, idealization or simplification, that are used in deterministic hydrology for rainfall calculations.

6.2 Which of the following sentences imply fuzziness in hydrological sciences?
   (a) Areal average rainfall distribution is constant.
   (b) Infiltration is rather high and leads to deep percolation.
   (c) Snow melts late in spring due to climate change.

6.3 Order the following topics in ascending sequence of fuzziness:
   (a) "geology"
   (b) "meteorology"
   (c) "hydrology"
   (d) "hydrogeology"
   (e) "mathematics"

6.4 If the "temperature, "wind speed," and "humidity" input linguistic variables have 3, 4, and 5 fuzzy sets, how many rules will be in the mechanical input rule base?

6.5 Fuzzify the following hydrological linguistic variables on their given variation domain and consider at least four membership functions with proper fuzzy names:
   (a) Rainfall "intensity," [2–20] mm/sec
   (b) Aquifer "potentiality," [0–500] m^2/day
   (c) Water "quality," [100–2000] micro-mhos
   (d) "Porosity," [0–100].

6.6 Write the following composite crisp logic rule into a set of rules by considering separately "ANDing" and "ORing" logical words:

   (Ash-Shamiyah)AND(Al-Yamaniyah)OR(Bani Umayr)AND(Alaf)
   AND(Majarish)OR(Ash-Sharah)AND(Ar'ar)AND(Na'man)AND(Rahjan)
   OR(Ash-Shamiyah)AND(Al-Yamaniyah)AND(Bani Umayr)
   AND(Majarish)OR(Ash-Sharah)AND(Ar'ar)AND(Na'man)

6.7 Assuming that all numbers are dams, determine all the crisp logical statements for water to flow from point A to point B in the following figure. (Assume that any dam is operating if the key is "close;" otherwise it is "open").

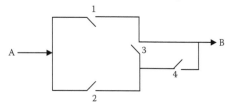

6.8 Two input linguistic variables are "*U*" and "*V*" and each one has three fuzzy sets for the description of their variation domain. Determine the mechanical input rule base and arrive at the complete rule base given the following scatter diagram as actual data:

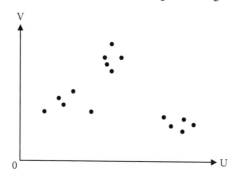

6.9 Given the two rules from a rule base as in the following, identify the numerical range of data for firing these two rules.

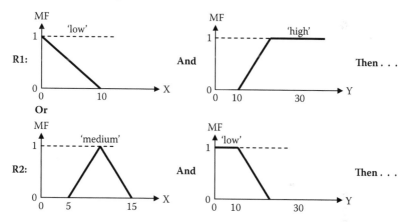

6.10 If a process has the following rules in the complete rule base then interpret the generating mechanism of this event.

"IF *X* is "small" AND *Y*'s 'big' THEN *Z* is 'low.'"

"IF *X* is "small" AND *Y*'s 'medium' THEN *Z* is 'medium.'"

"IF *X* is "medium" AND *Y*'s 'big' THEN *Z* is 'high.'"

"IF *X* is "big" AND *Y*'s 'big' THEN *Z* is 'big.'"

"IF *X* is "small" AND *Y*'s 'small' THEN *Z* is 'low.'"

6.11 Two input linguistic variables are given with the following membership functions, where "L," "M," "H," "S," and "B" correspond to "low," "medium," "high," "small," and "big" descriptions.

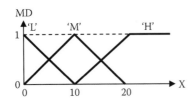 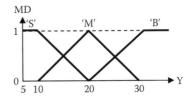

(a) Write the triggering rules for data as $X = \{2.5, 8.3, 12.4, 21.7, 25.3\}$ and $Y = \{4.5, 11.3, 16.9, 7.5, 29.8\}$.

(b) Write the triggering rules if the data has the scatter diagram shown in the following figure:

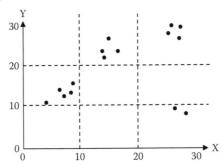

6.12 Find any mistakes in the following linguistic variables that are input variables for some hydrological modeling, and in the case of mistakes, suggest the right membership functions with reasons.

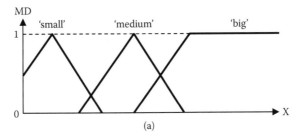

(a)

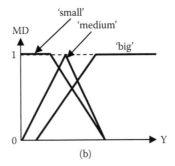

(b)

6.13 In the following table, triggering of rules with three input linguistic variable and one output variable in the case of 17 data values are given:

Data Number	Triggered Rule
1	R1; R5
2	R3; R2; R8
3	R2; R7
4	R4; R7
5	R5; R6
6	R2; R6; R8
7	R1; R3; R5
8	R6; R7
9	R3; R8
10	R5; R8
11	R3; R5; R7
12	R2; R4; R6
13	R4; R8
14	R3; R8
15	R2; R7
16	R1; R3; R5
17	R3; R2; R8

(a) Calculate the degree of belief for each rule.

(b) How many distinctive groups are there in this data?

6.14 The complete rule base of a hydrological event is given as follows:

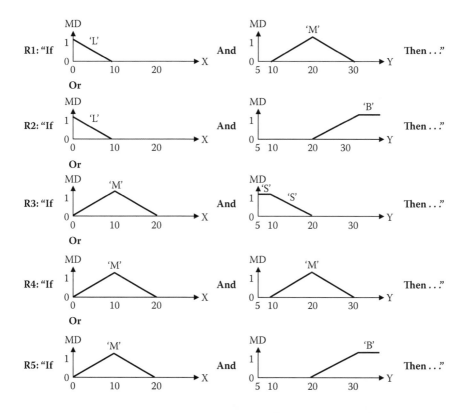

(a) Suggest five data values that fire R3 only.
(b) Suggest three data values that trigger R2, R4, and R5.
(c) Is it possible to have data values that trigger all the rules in this complete rule base?

6.15 Write the rule base from the following association relationships:

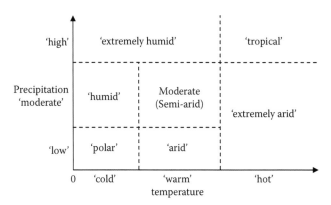

7 FIS
Fuzzy Inference Systems

7.1 GENERAL

Fuzzy systems are structured numerical estimators. They start from highly formalized insights about the structure of categories found in real life and then articulate fuzzy "IF..,...,...THEN...,...,..." rules as a kind of expert knowledge. Fuzzy systems combine fuzzy sets with fuzzy rules to produce overall complex nonlinear behaviors. The traditional hydrologic systems are so well developed, why bother with fuzzy models? In many cases, the mathematical model of the prediction process may not exist, or may be too expensive in terms of computer processing power and memory, and thus a system based on empirical rules may be more effective. Furthermore, FL is well suited to low-cost implementations based on cheap preliminary studies. Such systems can be easily upgraded by adding new rules to improve performance or add new features. In many cases, fuzzy inference systems (FIS) can be used to improve existing traditional systems by adding an extra layer of intelligence to the current models. Fuzzy system design is based on empirical methods, basically a methodical approach to trial and error.

There are different models for the estimation, prediction, and control toward better operation and management of water resources planning and maintenance. In the past, such goals were achieved through crisp mathematical statements and the use of digital computers for the interim calculations. Classical applications did not enhance the hydrologist with philosophical thinking and useful background logical deductions but mechanical application of ready formulations without attention to the underlying set of assumptions, approximations, idealizations, simplifications, and logical foundations. Perhaps this is one of the main reasons for the significant differences in opinion between operators, engineers, and academicians who cannot agree on scientific issues because each one has a different approach or interpretation of the same problem. For instance, most academicians have a formula addiction, operators prefer rule-of-thumb solutions dependent on many years of experience, and engineers prefer rather simple and readily applicable formulations. However, without any career distinction, those interested in the hydrological sciences may have the same common basis in terms of rationality and logical thinking problem solving although they may differ in their philosophical approach. They can ultimately agree on common logical conclusions linguistically. In this context, FL is a problem-solving methodology that lends itself to implementation in systems ranging from simple, small, embedded micro-scale controllers and predictors to large, networked, multichannel personal computer- or workstation-based data acquisition and control systems. It can be implemented in hardware, software, or a combination of the two. Fuzzy systems provide a simple way to arrive at a definite conclusion based on vague, ambiguous, imprecise, noisy, or missing input information.

Human processing of information is not based on CL, but rather fuzzy perceptions, fuzzy truths, fuzzy inferences, all resulting in an averaged, summarized, normalized output, to which a human assigns a precise number or decision value as he/she verbalizes it, writes it down, or acts on it. It is the FL modeling goal to do this by fuzzy inference systems (FIS). Hence, the primary focus of this chapter is to provide FIS model approaches for problem solving in water resources with linguistic statements in addition to numerical data. In particular, after having developed the rule base as explained in Chapter 6, it is now time to develop mechanisms for output estimation, prediction, and/or control purposes.

7.2 FUZZY INFERENCE SYSTEMS (FIS)

A fuzzy expert system with its FIS uses a collection of fuzzy membership functions (MFs) and rules—instead of mathematical statements—provide approximate reasoning about data where the rules are usually similar to the explanations given in Chapter 6. Any expert system has two major functions: namely, the problem-solving function capable of using variation domain-specific knowledge, and the user interaction function, which includes an explanation of the system's intention during and after the problem-solving process. The expert system is often expected to be able to deal with uncertain and incomplete information. In general, an expert system is a user-interactive setup, as shown in Figure 7.1, which consists of three major parts:

1. *Knowledge base:* This part includes knowledge that is specific to the application domain with facts about the domain and rules that describe relations in the domain. Herein, "IF…,…,…THEN…,…," rules are by far the most popular formalism for presenting knowledge.
2. *Interference engine:* This uses the knowledge actively in order to perform reasoning to obtain answers for users' queries.
3. *User interface:* This part provides smooth communication between the user and the system, further providing the user with insight into the problem-solving process through the inference.

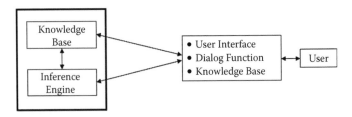

FIGURE 7.1 An expert structure.

The usefulness of an expert system occurs in dealing with imprecision in order for it to be a useful tool. Fuzzy expert systems incorporate FL and sets into the reasoning process and/or knowledge representation.

As the complexity of the system increases, it becomes increasingly difficult to formulate a mathematical model. The fuzzy model replaces the role of the mathematical model with another one that is built from a number of smaller rules. In general, they describe only a small section of the whole system. The FIS binds them together to produce the desired outputs. Fuzzy systems store banks of fuzzy associations or common-sense rules such as that which might be articulated by a human expert. Some rainfall occurrences are "heavier" than others so that the single fuzzy association ("heavy," "more") encodes all these combinations. Fuzzy systems directly encode structural knowledge but in a numerical framework by entering the fuzzy association ("heavy," "more space") as a single entry in a rule base.

Experts in the water sciences can sketch quickly the approximate shape of relevant fuzzy sets for the problem at hand. After the hydrologist runs the model and examines its consequences, the precise characteristics of the fuzzy vocabulary can be adjusted, if necessary. Establishment of fuzzy sets and building an FIS are "faster" and "quicker" than conventional knowledge-based systems using crisp constructs and numerical data only. FIS routinely shows a one- or two-order of magnitude reduction in rules because it simultaneously handles all the interlocking degrees of freedom. FISs are "very robust" because the overlapping of the fuzzy regions, representing the continuous domain of each cause-and-effect variable, contributes to a well-behaved and predictable system operation. These systems are validated in the same manner as conventional systems. The tuning of FISs, however, is usually "much simpler" because there are "fewer rules."

Another common way of understanding FL is to look at quantity magnitudes. If a standard mathematical logic system is used, then it is necessary to have a crisp cutoff number.

Conceptually, FISs are "very simple" because they consist of input and output stages with a processing stage in between. The input stage maps the input variable to the appropriate MFs, as previously explained in Chapter 6. The processing stage invokes appropriate rules and generates a result for each, and then combines the results of the rules. Finally, the output stage converts the combined result back into a specific output value.

The use of fuzzy set theory allows the user to include the unavoidable imprecision in the data. FIS is the actual process of mapping from a given set of input variables to an output based on a set of fuzzy rules, which are the essence of the modeling. The following steps are necessary for the successful application of a fuzzy model:

1. The input and output variables are fuzzified by considering convenient linguistic subsets such as "high," "medium," "low," "heavy," "light," "hot," "warm," "big," "small," etc.
2. Rules are constructed based on expert knowledge and/or available literature. The rules relate the combined linguistic subsets of input variables to the convenient linguistic subset.

3. The implication part of an FIS is defined as the shaping of the consequent predicate based on the premise (antecedent) part and the input is a fuzzy set.
4. The result appears as a fuzzy set, and therefore it is necessary to defuzzify the set to a crisp value that can be used by the administrator or engineer. There are various defuzzification procedures in the applications.

The general steps in any FIS application in practice are included in Figure 7.2.

Fuzzy sets, "IF…THEN…" rules, and reasoning concepts constitute the basis of the soft computing framework as in Chapter 6, which is referred to as the FIS and known under a variety of names such as fuzzy models, fuzzy-rule-based systems, fuzzy expert systems, fuzzy associative memory, fuzzy logic controller, and, simply and ambiguously, fuzzy systems. The fundamental structure of FIS includes three interconnected components: a database, a rule base, and a reasoning mechanism.

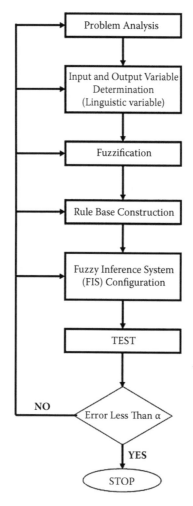

FIGURE 7.2 FIS process steps.

The database and rule base were the focus of Chapter 6. The linguistic database defines the MFs that will be used in the rule base. Hence, it can also be considered a dictionary that includes fuzzy wordings. The reasoning mechanism performs the inference procedure based on fuzzy reasoning such that given rules and facts about the concerned problem, reasonable outputs and conclusions are derived and interpreted.

7.3   MAMDANI FIS

This FIS is completely fuzzy in the input, output, and rule base, and hence the estimations are fuzzy sets that are not formal but in different shapes without any MD equal to 1. The use of fuzzy set theory allows the user to include the unavoidable imprecision in the data. Mamdani (1974) FIS is the actual process of mapping from a given set of input variables to an output based on a set of fuzzy rules with fuzzy inputs and outputs. The essence of the modeling is the setup of the fuzzy rules, which help for inference. The following steps are necessary for the successful application of a Mamdani FIS. The general fuzzy inference engine proceeds in four successive steps:

1. *Fuzzification:* The MFs defined in the input variables are applied to their actual values to determine the MD for each rule premise.
2. *Inference:* The MD for the premise of each rule is computed and applied to the conclusion part of each rule. This results in one fuzzy subset to be assigned to each output variable for each rule. According to either "minimum," "ANDing," or "product" conjunctives, the output MF is clipped off at a height corresponding to the rule premise's computed MD. In the "product" inference, the rule premise's computed MD scales the output MF.
3. *Composition:* All the fuzzy subsets assigned to each output variable are combined together to form a single fuzzy subset for each output variable. Again, usually "maximum" or "summation" is used. In "maximum" composition, the combined output fuzzy subset is constructed by taking the pointwise maximum over all the fuzzy subsets assigned to a variable by the inference rule, which corresponds to FL "ORing." In "summation" composition, the combined output fuzzy subset is constructed by taking the pointwise sum over all the fuzzy subsets assigned to the output variable by the inference rule.
4. *Defuzzification:* This is used to convert the fuzzy output set to a crisp number. There are different defuzzification methods that can be used according to the overall purpose, as explained later in this chapter. The general framework of the FIS structure is presented in Figure 7.3.

The first application was proposed by Mamdani (1974) as a set of linguistic control rules based on human operation experience for controlling a steam engine and boiler combination. Figure 7.4 indicates how a two-rule Mamdani FIS derives the overall output "*Y*" from the "*X*" input variable. The deduction appears as a non-normal fuzzy set, which is then converted to a crisp value according to one of the defuzzification methods.

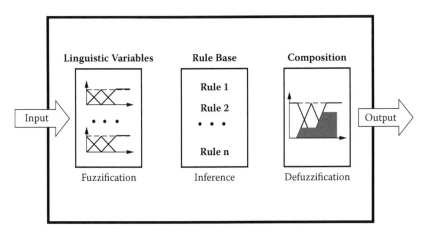

FIGURE 7.3 General FIS structure.

Mamdani and Assilian (1975) were the first researchers to apply fuzzy approximate reasoning by "max-min" composition as shown in Figure 7.5, which is the model for the intersection of fuzzy sets.

In Mamdani FIS, there are two similar inference systems with either crisp or fuzzy inputs. First, the crisp input Mamdani FIS is explained, wherein the first step in such an approach necessitates the rule base with its most explicit form as given in Figure 7.6 with two inputs and a single output.

It is possible to show all the input and output MFs on the same graph and then, accordingly, the crisp data values fire the convenient fuzzy sets. Such an example is shown in Figure 7.6, with two inputs and one output variable. In this manner, possible MFs are shown collectively. This figure yields non-normal fuzzy outputs according to "min-max" inference system outputs. In the same figure, the complete rule base is already prepared according to the explanations in Chapter 6 and then the data values are entered on the horizontal axis as the X_1 and Y_1 data pair and two rules are triggered. Because each linguistic variable fuzzy set is related to another one through "ANDing" according to fuzzy set operators, it is necessary to take the

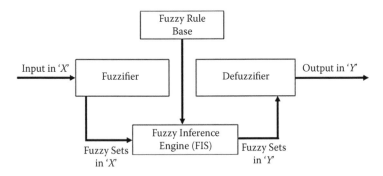

FIGURE 7.4 General fuzzy system.

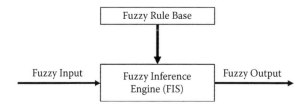

FIGURE 7.5 Pure fuzzy system.

minimum MD among the two, which are d_1 and d_2 in the first rule and d_3 and d_4 in the other. "ANDing" means minimization and therefore the minimum MDs are transferred to the output section of the rule base. Accordingly, the transferable MDs are d_2 and d_4, respectively. Transition of d_2 on the consequent part of the rule now depicts MD on the vertical axis and then onward the corresponding output fuzzy set is truncated into two parts, as triangle above and trapezium below. Because the minimization of the MDs is adopted, the truncated trapezium is considered the output fuzzy set on the consequent part of the first rule. It is shown as shadow in Figure 7.6. This trapezium is not normal but convex in the sense that there is no element with MD equal to 1. Hence, in the inference systems, the consequent part fuzzy sets are not standardized to have the formal fuzzy set property. Similarly, for

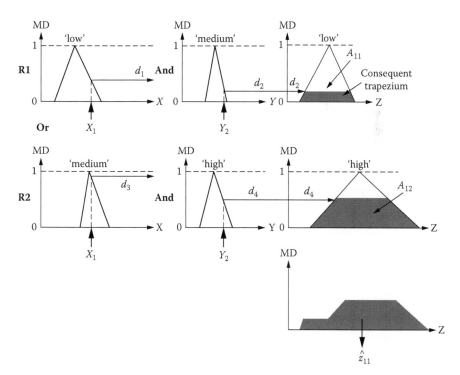

FIGURE 7.6 Min-max Mamdani FIS with crisp data values.

the second rule at the d_4 MD level, there is also another trapezium but it does not have the same base. The first one is the remnant of "low" whereas the other one is the "high" fuzzy set truncations. Finally, the consequent part inferences are available individually for each rule.

After this operation, the antecedent parts of each rule are not needed anymore and now the question becomes how to combine together the triggered rules' consequent parts. The rules are connected by "ORing" in the rule base; and in fuzzy set terminology, it corresponds to the maximization of the consequent parts (see Chapter 3). Maximization of the two consequent truncated trapeziums leads to overlapping of the two, as shown on the bottom right-hand side in Figure 7.6. If there were more than two rules fired, then the final combination of the consequent parts will appear in an irregular shape of non-normal fuzzy set, which is not convex either. This inference system is referred to as the "min-max" operation because first minimization ("ANDing") and then maximization ("ORing") operations are applied. It is also possible to show all the fuzzy sets collectively as in Figure 7.7.

The "min-max" procedure ignores all the consequent part fuzzy sets except the one with the minimum firing MD. This means that the minimization operation does not take into consideration the contribution from each fired fuzzy set in the antecedent parts of the rules. Alternatively, the product of the fired MDs in the antecedent part can be adopted as the combined MD for the interference on the consequent part as shown in Figure 7.8. The interference of each rule on the consequent part is considered a triangle with height equal to the product of antecedent MDs. Hence, as in Figure 7.8, the product of MDs is $d_1 d_2$ and in the second rule, $d_3 d_4$. Subsequently, the inter-rule interference operation in this FIS is referred to as the "prod-max" procedure. Because the MDs from each input linguistic variable fuzzy set are less than 1, their multiplication results in almost zero in the case of more than four or five input variables. Hence, although the contribution of each input variable fuzzy set is taken

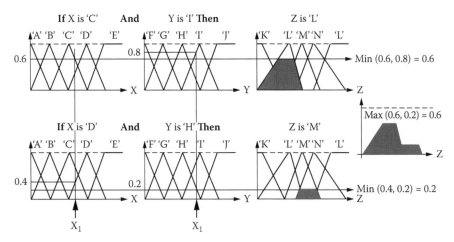

FIGURE 7.7 Mamdani inference with collective MFs and crisp input data.

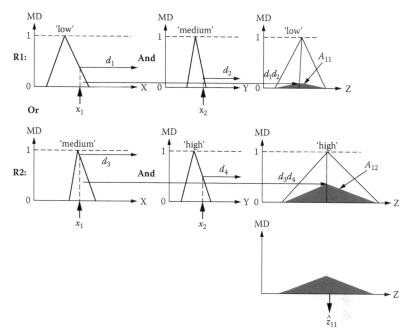

FIGURE 7.8 Min-max Mamdani FIS with crisp fuzzy values.

into consideration, their multiplication might result in a small value that may not be representative of the real situation. To be on the safe side, most often the "min-max" procedure is applied in the practical applications. Figure 7.9 indicates all the rules in a more compact form with inferences.

On the other hand, it is possible that the data is not crisp but in the form of fuzzy numbers, and in this case the Mamdani FIS takes the rule base form as given in Figure 7.10. Herein, the intersection of the fuzzy numbers with the fuzzy sets is considered the MD and then according to "min-max" procedure, the inference is shown in the same figure, where the dotted triangles are the fuzzy numbers on the antecedent part of each rule.

In Figures 7.6 through 7.10, the final product appears in the form of a composite fuzzy set with all MDs less than 1, and the final crisp solution should be sought from this set. In practical applications, it is necessary to obtain a single crisp value from of this composite fuzzy shape through a defuzzification operation.

7.4 DEFUZZIFICATION

Defuzzification is the reverse process of fuzzification. It converts the confidences in a fuzzy set of word descriptors into a real number. It is necessary when the output is required as a crisp number by the user. For example, in a runoff estimation expert system, it may be necessary to tell the user how many of a certain item are expected to be in a cubic meter per second (m^3/sec).

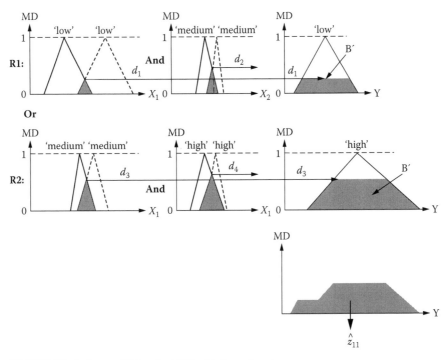

FIGURE 7.9 "min-max" Mamdani FIS with fuzzy inputs.

Although there are different ways of defuzzification, in fuzzy reasoning, quite simple methods are used. It is intuitive that fuzzification and defuzzification should be reversible. That is, if a number is fuzzified into a fuzzy set, when it is immediately defuzzified one should get the same number back again. If defuzzification

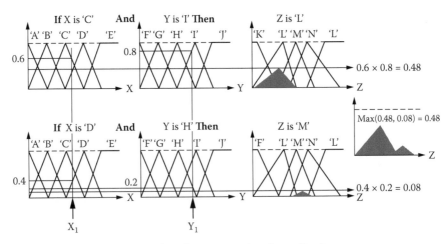

FIGURE 7.10 The "prod-max" interference procedure in a collective manner.

is to take place, this has implications for the shape of MFs used to fuzzify input variables into fuzzy sets used in the reasoning process, which ultimately results in the defuzzification of an output fuzzy set. Although there are many defuzzification procedures available for fuzzy control, fuzzy reasoning applications can usually be satisfied with fewer options. The best shape for a MF depends on its use. If the ultimate output is nonnumeric, flat-topped MFs with adjacent functions having maximum overlap are usually the best. On the other hand, if numeric output is desired, peaked functions intersecting at half-full confidence are usually the easiest to manage.

The results appear as irregular shapes (Figures 7.6 through 7.10) and the final step in the approximate reasoning algorithm is defuzzification—that is, choosing one crisp value for the output variable. The fuzzy reasoning algorithm using the Mamdani model leads to an output fuzzy set with particular MDs of possible numerical values of the output variable. Defuzzification compresses this information as an output value. If a certain defuzzification procedure is adopted for the calculations, then the same procedure must be kept throughout the estimation, prediction, or control processes. Changing the defuzzification procedure within the same problem causes problems because then a haphazard defuzzification is applied, which cannot satisfy the standard approach.

7.4.1 ARITHMETIC AVERAGE

The arithmetic average is a procedure that gives equal weight to each fuzzy output and the fuzzy set is sampled at regular intervals, say n points as in Figure 7.11, and then the arithmetic average of all the values is taken into consideration. For this purpose, the minimum Z_m and maximum Z values on the horizontal axis are calculated; and, as in histogram analysis in classical statistics, the range is divided into sub-classes and each class boundary value is considered in the calculation. Finally, two sequences—one for the output values and the other one for the corresponding MDs—are obtained as $Z_m, Z_1, Z_2, \ldots, Z_M$ and $0\ d_1\ d_2 \ldots d_k\ 0$, respectively.

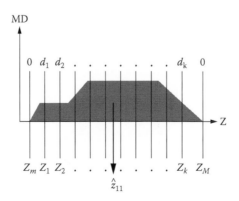

FIGURE 7.11 Output fuzzy set intervals.

The arithmetic average as defuzzification can be obtained from:

$$\hat{z}_{11} = \frac{1}{k+2}\left[\sum_{i=1}^{k} Z_i + Z_m + Z_M\right] \tag{7.1}$$

7.4.2 Weighted Average

There are various weighted average defuzzification methods used in practical applications. The first one is based on the equal interval division as in Figure 7.11 with the following formulation:

$$\hat{z}_{11} = \frac{\displaystyle\sum_{i=1}^{k} d_i Z_i}{\displaystyle\sum_{i=1}^{k} d_i} \tag{7.2}$$

Each rule has the centroid of the consequent part value, as shown in Figure 7.12.

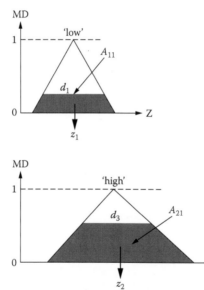

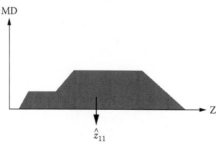

FIGURE 7.12 Detailed output fuzzy sets.

It is possible to consider the weights as the areas of the rule consequent part fuzzy sets and hence the defuzzification takes the following form:

$$\hat{z}_{11} = \frac{\sum\limits_{i=1}^{k} A_{i1} z_i}{\sum\limits_{i=1}^{k} A_{i1}} \qquad (7.3)$$

where k is now the number of triggered rules. Finally, it is possible to calculate the defuzzified compound output by considering the MDs as the weighting factors, which leads to:

$$\hat{z}_{11} = \frac{\sum\limits_{i=1}^{k} d_i z_i}{\sum\limits_{i=1}^{k} d_i} \qquad (7.4)$$

7.4.3 CENTER OF GRAVITY (CENTROID)

This is the most frequently used fuzzification procedure because it has more physical appeal than any other procedure (Lee, 1990). Its mathematical expression is given as:

$$\hat{Z} = \frac{\int \mu(z) z \, dz}{\int \mu(z) \, dz} \qquad (7.5)$$

The result obtained this method is shown in Figure 7.13.

7.4.4 SMALLEST OF MAXIMA

This method is useful in the case of rare event modeling such as droughts, which are among the smallest group of the data or future possible data values (Figure 7.14). First, the maximum MD range is considered, which is shown in the figure by R. All the values within this range have the same MDs, and the most convenient for the problem at hand may be the smallest of the maxima.

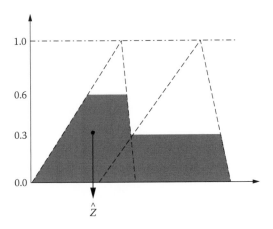

FIGURE 7.13 Center of area method defuzzification.

7.4.5 Largest of Maxima

This is very similar to the previous method. The only difference is that the maximum Z value is taken from the maximum MD range (Figure 7.15).

7.4.6 Mean of the Range of Maxima

This method is also called middle of the maximum and it has the same definition of the range as in the two previous sections. If the lower and upper values of the maximum range are Z_L and Z_U, respectively, then this method yields defuzzification value as:

$$\hat{Z} = \frac{Z_L + Z_U}{2} \tag{7.6}$$

The corresponding shape is given in Figure 7.16.

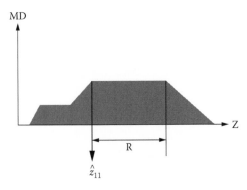

FIGURE 7.14 Smallest of maxima.

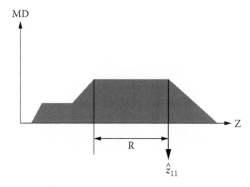

FIGURE 7.15 Largest of maxima.

7.4.7 LOCAL MEAN OF MAXIMA

This is a more generalized method of the mean of maxima defuzzification procedure. All maximum MD ranges in the FIS output are considered as in Figure 7.17.
After finding the means of (Z_1 and Z_2) of the ranges of separate maxima regions, they
are weighted with MDs d_1 and d_2 as:

$$\hat{Z} = \frac{d_1 Z_1 + d_2 Z_2}{d_1 + d_2} \tag{7.7}$$

In most the applications, it is advised that the hydrologist should look at the MDs
of individual variable and choose one of the defuzzification methods according to

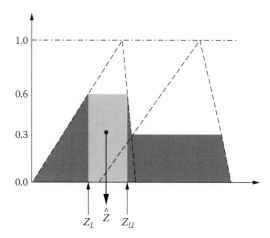

FIGURE 7.16 Mean of maxima method defuzzification.

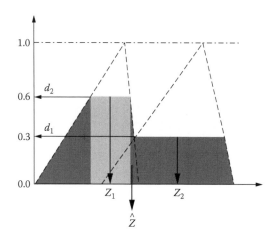

FIGURE 7.17 Local mean of maxima method defuzzification.

the purpose of the problem. One can even make one's own preference according to the problem at hand.

Defuzzification causes a great deal of information loss. Therefore, in any hydrologic design, it is recommended that the hydrologist take into consideration some other features of the final composition fuzzy set in addition to formal defuzzification methods.

7.5   SUGENO FIS

This model is also known as Takagi-Sugeno-Kang (TSK) FIS according to its proposers' initials (Sugeno, 1985; Takagi and Sugeno, 1985. They devised a systematic approach for generating fuzzy rules from a given set of input-output data. The rule base generated by considering only the antecedent parts of the rules remains the same but the difference comes in attaching consequent parts, which are no longer in the form of fuzzy sets but simple linear mathematical functions. In a two-input ("X" and "Y") and single-output ("Z") domain, a typical rule takes the following form:

R: "IF X is 'small' AND Y is 'large' THEN $Z = f(X, Y)$"

where $f(X, Y)$ is a crisp function that constitutes the consequent part of the fuzzy rule. Usually, it has a linear form in terms of input variables X and Y. Its representative form can be obtained from a given set of data, or it can also be any function as long as it can appropriately describe the output of the model within the fuzzy region specified by the antecedent part of the rule base. In practical works, two alternatives of $f(X, Y)$ are used. One alternative has a constant (c)

consequent part as

R: "IF X is 'small' AND Y is 'large' THEN $Z = c$"

which is also referred to as a zero-order TSK or Sugeno FIS. The other alternative has a linear consequent part known as a first-order Sugeno FIS:

R: "IF X is 'small' AND Y is 'large' THEN $Z = c + aX + bY$."

If there are more than two variables, then the consequent part of each rule base appears similar to a linear multiple regression expression. This alternative is similar to a linear interpolation procedure, the only difference being that the interpolated point has some MD depending on the aggregation of the input variables through the fuzzy word MFs in the antecedent part of the rules.

The zero-order Sugeno FIS yields a smooth function of its input variables as long as the neighboring MFs have enough overlap. Even in the Mamdani FIS, the smoothness of the output can be obtained through the overlapping of MFs in the antecedent part, whereas the consequent part does not play any role in such a smoothing procedure. It is therefore possible to adjust the smoothness of the final functional form by playing with the antecedent part MFs either through "trial and error" or by a "systematic" approach such as adaptive-network-based FIS (ANFIS), which includes neural networks or genetic algorithms (Şen, 2004a, b).

In the Sugeno FIS, each rule has a crisp output and hence the overall output is obtained as a weighted average, which avoids the time-consuming procedure of defuzzification as required by Mamdani FIS (Figure 7.18).

On the other hand, the Sugeno FIS structure can be presented as in Figure 7.19, where the output appears in the form of a weighted average as:

$$\hat{Z} = \frac{\sum_{i=1}^{n} w_i f_i}{\sum_{i=1}^{n} w_i} = \sum_{i=1}^{n} \alpha_i f_i \tag{7.8}$$

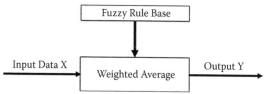

FIGURE 7.18 Partial fuzzy system.

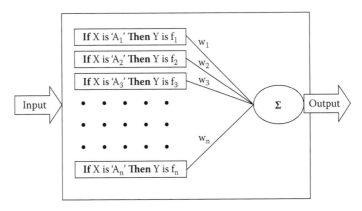

FIGURE 7.19 Sugeno FIS structure.

Example 7.1

In Table 7.1, the "X" and "Y" linguistic variables are given and the purpose is to develop a Sugeno FIS model that depicts the line defined by these points. Such a model provides "Y" estimations for any given value of "X." First, the scatter of data is presented in Figure 7.20.

Because Sugeno FIS will be established in the case of such small data numbers, it is possible to take the number of input MFs equal to the data number, and each MF will be adopted as triangular or trapezium with the maximum MD at the triangular or trapezium apex and the base lies between two successive "X" values. Such a configuration of MFs is given in Figure 7.21. First, the input linguistic variable is fuzzified as in this figure with five MFs as fuzzy sets labeled "X1," "X2," "X3," "X4," and "X5."

By considering Y values and the MFs in Figure 7.21, one can write the necessary rules in a rule base of the convenient Sugeno FIS as:

R1: "IF X is 'X1' THEN $Y_1 = 37.6$."

R2: "IF X is 'X2' THEN $Y_1 = 28.5$."

R3: "IF X is 'X3' THEN $Y_1 = 8.4$."

TABLE 7.1
Data

Data Number	X	Y
1	10	37.6
2	35	28.5
3	50	8.4
4	95	21.3
5	140	25.2

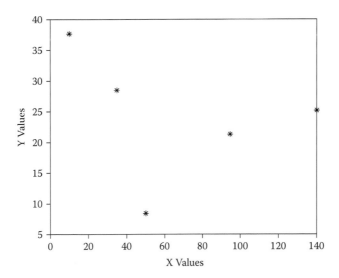

FIGURE 7.20 Scatter diagram.

R4: "IF X is 'X4' THEN Y_1 = 21.3."

R5: "IF X is 'X5' THEN Y_1 = 25.2."

According to this rule base, any given X value will trigger at least two of the rules and accordingly each one with an MD, and these MDs will act as weights and hence the final result from the triggered rules will be calculated from Equation (7.8). For instance, if X = 15 is the input value as shown in Figure 7.21, then the MDs corresponding to "X1" and "X2" fuzzy sets are 0.20 and 0.80, respectively.

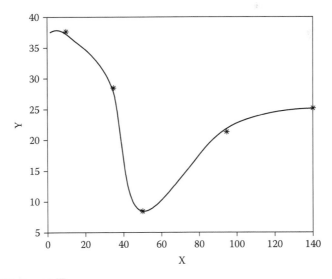

FIGURE 7.21 Input MFs.

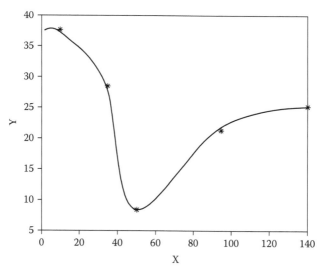

FIGURE 7.22 Sugeno FIS model results.

These mean that $X = 15$ triggered R1 and R2 and consequently the final Y estimation from Equation (7.8) is obtained as 35.78. Likewise, selection of a set of X values within the variability domain results in corresponding Y values and, finally, the curve that connects these points appears as in Figure 7.22.

7.6 TSUKAMOTO FIS

Here, the consequent of each fuzzy "IF...THEN..." rule is presented by a fuzzy set with a monotonic MF, as in Figure 7.23 (Tsukamoto, 1979). In this manner, after the transition from the antecedent part to the consequent, each rule output value can be

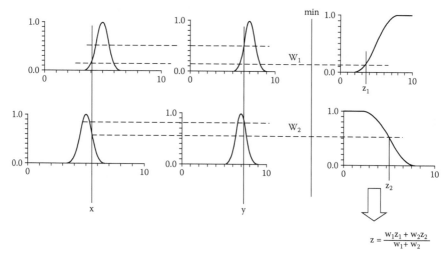

FIGURE 7.23 The Tsukamoto FIS.

known without any defuzzification directly from the consequent MFs, which either have an S-shape or a Z-shape. As explained in Chapter 3, different CDF type MFs can also be adopted for the consequent part.

Each rule fires a crisp value from the corresponding monotonic MF, depending on the strength of the rule aggregation procedure. The overall output is then taken as the weighted average of each rule's output (Equation 7.7). Again, Tsukamoto FIS avoids the time-consuming defuzzification procedure by using weighted average, but it is not used often because it is not as transparent as either Mamdani or Sugeno FIS.

7.7 ŞEN FIS

This inference system was developed by S, en (1998) for solar irradiation estimation. Herein, input and output fuzzy MFs are considered collectively first without any rule base (Figures 7.24 and 7.25). Entry of the input data i_1 into Figure 7.24 leads to the triggering of two MFs with respective MDs of α and β, which correspond to I_3 and I_2 input MFs, respectively. The same α and β values are entered on the MD axis of the output MFs as in Figure 7.25. At the end, four readings from the horizontal output axis are obtained as (O_{31}, O_{32}) and (O_{41}, O_{42}) from output MFs "O_3" and "O_4," respectively. The output MFs are depicted according to the relevant rule base. The arithmetic average of these output couples is calculated as crisp values:

$$\overline{O}_3 = \frac{O_{31} + O_{32}}{2}$$

and

$$\overline{O}_4 = \frac{O_{41} + O_{42}}{2}$$

The final output value $\hat{O}_1$ is the weighted average of these two values, which is calculated as follows:

$$\hat{O}_1 = \alpha\overline{O}_3 + \beta\overline{O}_4 \tag{7.9}$$

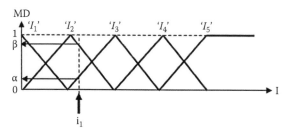

FIGURE 7.24 Input MFs.

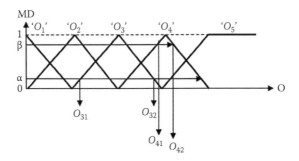

FIGURE 7.25 Output MFs.

7.8 FIS TRAINING

Up to this point the rule base and FISs operated according to the initial fuzzy set attachments to the input and output variables, without changing their positions or shapes according to the available data set. However, it is necessary to train the rule base by applying the convenient FIS. Figure 7.26 shows that changing one or more fuzzy sets in the antecedent part of the rule base will cause a difference in the final estimation no matter what the FIS is. Comparison of this figure with Figure 7.6 indicates that at least three fuzzy sets in the antecedent part are changed, as shown by the broken lines; and, accordingly, the minimization in R1 now leads to d_1 being smaller than d_2. Whereas in the second rule, the same MD remains as the minimum.

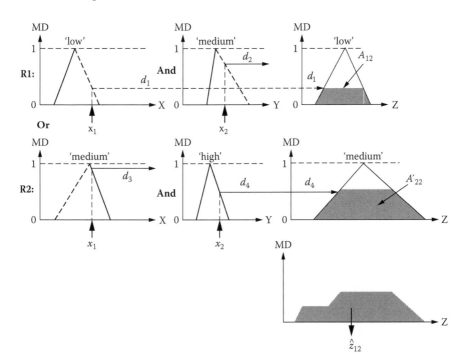

FIGURE 7.26 Training with fuzzy sets.

Hence, the consequent part of the first rule changes and the overall inference, after maximization of the consequent parts, also changes. In fact, there is an increase in the final defuzzified value. Any change in the position of any MF in the antecedent part will cause a change in the consequent, and subsequently in the overall interference after defuzzification and in the final crisp result.

This procedure shows that by playing with the antecedent part of the fuzzy sets (MFs), it is possible to make the final defuzzified value very close to the measurement and that it is even possible to reduce the error to zero. If this is the case, then in the case of triggering the same rule again by any other data during its training, the fuzzy sets will change and hence zero error in the previous data value will be demolished. This point indicates that rather than making the error for a particular data equal to zero, it is necessary to set the overall data from all the data values as the minimum. For this purpose, the two criteria are:

1. The summation of the errors should be equal to zero practically.
2. Additionally, the summation of the data error sum of squares must be minimum, which corresponds to the least squares error criterion.

7.9   TRIPLE VARIABLE FUZZY SYSTEMS

Assume that the variables x, y, and z all take on values in the interval $[0, 10]$ with two MFs as:

$$\text{"low"} = 1 - (t/10)$$

and

$$\text{"high"} = t/10$$

where t is a dummy variable that represents either x or y or z. The graphical representation of these fuzzy subsets is given in Figure 7.27. Let the following set of rules be valid between the input (x and y) and output (z) variables. This is similar to three-dimensional regression methodologies.

R1: "IF x is 'low' AND y is 'low' THEN z is 'high.'"

R2: "IF x is 'low' AND y is 'high' THEN z is 'low.'"

R3: "IF x is 'high' AND y is 'low' THEN z is 'low.'"

R4: "IF x is 'high' AND y is 'high' THEN z is 'high.'"

According to Mamdani FIS, the following points are noteworthy:

1. In this example, MDs of "low" and "high" add up to 1 for all t. This is not required but it is fairly common

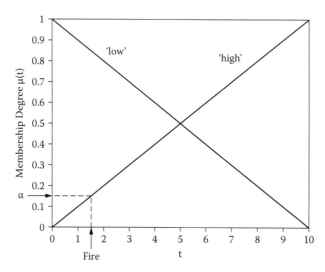

FIGURE 7.27 MFs.

2. The value of t at which "low" is maximum is the same as the value of t at which "high" is minimum, and vice versa. This is also not required but it is fairly common.
3. The same MFs are used for all variables. This is not required and is also not common.

In the fuzzification process, the MFs defined on the input variables are applied to their actual values to determine the degree of truth for each rule premise. The degree of truth for a rule's premise is sometimes referred to as its α value. As already mentioned, if a rule's premise has a non-zero degree of truth (if the rule applies at all), then the rule is said to fire or trigger.

Example 7.2

If the two input values (x and y) are given as in Table 7.2, then the MD for each fuzzy subset ("low," "high") and output α values can be calculated accordingly.

In the inference process, the MD for the premise of each rule is computed and applied to the conclusion part of each rule. This results in one fuzzy subset assigned to each output variable for each rule. "min-max" and "min-prod" are two inference methods or inference rules (as mentioned in Section 7.3). In minimum inference, the output MF is clipped off at a height corresponding to the rule premise's computed degree of truth. This corresponds to the traditional interpretation of the FL "ANDing" operation. In product inference, the output MF is scaled by the rule premise's computed degree of truth.

Let's look at R1 for $x = 0.0$ and $y = 3.2$. As shown in Table 7.2, the premise degree of truth works out to 0.68. For this rule, minimum inference will assign z

TABLE 7.2
Input Values Together with MD and α

Input		MD				Output			
		x		y					
x	y	"low"	"high"	"low"	"high"	α_1	α_2	α_3	α_4
0.0	0.0	1.0	0.0	1.0	0.0	1.0	0.0	0.0	0.0
0.0	3.2	1.0	0.0	0.68	0.32	0.68	0.32	0.0	0.0
0.0	6.1	1.0	0.0	0.39	0.61	0.39	0.61	0.0	0.0
0.0	10.0	1.0	0.0	0.0	1.0	0.0	1.0	0.0	0.0
3.2	0.0	0.68	0.32	1.0	0.0	0.68	0.0	0.32	0.0
6.1	0.0	0.39	0.61	1.0	0.0	0.39	0.0	0.61	0.0
10.0	0.0	0.0	1.0	1.0	0.0	0.0	0.0	1.0	0.0
3.2	3.1	0.68	0.32	0.69	0.31	0.68	0.31	0.32	0.31
3.2	3.3	0.68	0.32	0.67	0.33	0.67	0.33	0.32	0.32
10.0	10.0	0.0	1.0	0.0	1.0	0.	0.0	0.0	1.0

to the output variable MF as:

$$R1: z = \begin{cases} \dfrac{z}{10} & \text{if } z \le 6.8 \\ 0.68 & \text{if } z \ge 6.8 \end{cases}$$

For the same conditions, product inference will assign z to the fuzzy subset defined by the MF as:

$$R1: z = 0.68\,\text{high}(z) = 0.068z$$

Note that the terminology used here is slightly non-standard. In most texts, the term *inference method* is used to mean the combination of the things referred to separately here as "inference" and "composition," such terms as *min-max* inference and *sum-product* inference in the literature. They are the combination of "max" composition and "min" inference, and "product" inference, respectively. As in this book, one can also see the reverse terms *min-max* and *product-sum* which mean the same things but in the reverse order. It seems clearer to describe the two processes separately.

In the composition process, all the fuzzy subsets assigned to each output variable are combined to form a single fuzzy subset for each output variable.

In max composition, the combined output fuzzy subset is constructed by taking the pointwise maximum over all the fuzzy subsets assigned to the output variable by the inference rule. In sum composition, the combined output fuzzy subset is constructed by taking the pointwise sum over all the fuzzy subsets assigned to the output variable by the inference rule. Note that this can result in truth-values greater than 1. For example, assume $x = 0.0$ and $y = 3.2$; then the min inference would assign the following four fuzzy subsets to z:

$$R1: z = \begin{cases} \dfrac{z}{10} & \text{if} \quad z \leq 6.8 \\ 0.68 & \text{if} \quad z \geq 6.8 \end{cases}$$

$$R2: z = \begin{cases} 0.32 & \text{if} \quad z \leq 6.8 \\ 1 - \dfrac{z}{10} & \text{if} \quad z \geq 6.8 \end{cases}$$

R3: 0.0

R4: 0.0

Max composition would result in the fuzzy subset:

$$\text{fuzzy}(z) = \begin{cases} 0.32 & \text{if} \quad z \leq 3.2 \\ \dfrac{z}{10} & \text{if} \quad 0.32 \leq z \leq 6.8 \\ 0.68 & \text{if} \quad z \geq 6.8 \end{cases}$$

and product inference would assign the following four fuzzy subsets to z:

Rule 1: $0.068z$

Rule 2: $0.32 - 0.032z$

Rule 3: 0.0

Rule 4: 0.0

Sometimes, it is useful to examine just the fuzzy subsets that are the result of the composition process, but more often this fuzzy value must be converted to a single number, which is a crisp value. This is what the defuzzification process does. For example, using "min-max" inference and average-of-maxima defuzzification results in a crisp value of 8.4 for z.

7.10   ADAPTIVE NETWORK-BASED FIS (ANFIS)

Various FIS types have been studied in the literature and each one is characterized by consequent parameters. The adaptive-network-based FIS (ANFIS) model was proposed by Jang (1992) and applied successfully to many problems. ANFIS identifies a set of parameters through a hybrid learning rule that combines back-propagation gradient descent and a least squares method. It can be used as a basis for constructing a set of fuzzy "IF…THEN…" rules with appropriate MFs to generate the preliminary stipulated input-output pairs.

ANFIS applications and properties were investigated and several methods were proposed for partitioning the input space and hence address the structure identification problem. Fundamentally, ANFIS is a graphical network representation of Sugeno-type fuzzy systems, endowed with neural learning capabilities. The network is comprised of nodes with specific functions and waves, and are collected in layers with specific functions.

To illustrate ANFIS's representational strength, the neural fuzzy control system is considered based on the Sugeno fuzzy rules whose consequent parts are linear combinations of their preconditions. The Sugeno fuzzy rules are in the following forms:

$$R^j: \text{"IF } x_1 \text{ is } A_1^j \text{ AND } x_2 \text{ is } A_2^j \text{ AND}……\text{AND } x_n \text{ is } A_n^j$$

$$\text{THEN } f_j = a_0^j + a_1^j x_1 + a_2^j x_2 + \cdots + a_n^j x_n \text{"}$$

(7.10)

where x_i's $(i = 1, 2,…, n)$ are input variables, y is the output variable, A_i^j are linguistic words of the antecedent part with MFs $\mu_{A_i^j}(x_i)$, $(j = 1, 2,…, n)$, and $A_i^j \in R$ are coefficients of linear equations $f_i (x_1, x_2,…, x_n)$.

Assume that the fuzzy control system under consideration has two inputs x_1 and x_2, one output, and that the rule base contains two Sugeno fuzzy rules as follows (Figure 7.28a):

$$R^1: \text{"IF } x_1 \text{ is } A_1^1 \text{ AND } x_2 \text{ is } A_2^1 \text{ THEN } f_1 = a_0^1 + a_1^1 x_1 + a_2^1 x_2 \text{"}$$

(7.11)

$$R^2: \text{"IF } x_1 \text{ is } A_1^2 \text{ AND } x_2 \text{ is } A_2^2 \text{ THEN } f_2 = a_0^2 + a_1^2 x_1 + a_2^2 x_2 \text{"}$$

(7.12)

The fuzzification is inferred through weighted averaging in Equation (7.8), where w_j's are the firing strengths of R^i, $(j = 1, 2)$ and are given by:

$$\omega_j = \mu_{A_1^j}(x_1) + \mu_{A_2^j}(x_2) \qquad j = 1, 2.$$

(7.13)

If "product" inference is used, then the corresponding ANFIS architecture is shown in Figure 7.28b, where functions in the same layers are of the type described below. This is an artificial neural network architecture where the following meanings can be attached to each layer:

1. Layer 1: Every node in this layer is an input node that just passes external signals to the next layer.
2. Layer 2: Every node in this layer acts as an MF, $\mu_{A_i^j}(x_i)$, and its output specifies the degree to which the given x_i satisfies the quantifier A_i^j.

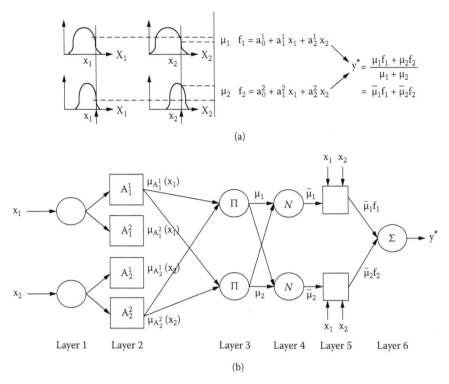

FIGURE 7.28 Structure of ANFIS: (a) FIS and (b) ANFIS.

Generally, $\mu_{A_i^j}(x_i)$ is selected as bell-shaped with a maximum equal to 1 and a minimum equal to 0, such as (Figure 28a):

$$\mu_{A_i^j}(x_i) = \frac{1}{1+\{[(x_i - m_i^j)/\sigma_i^j]^2\}^{b_i^j}} \tag{7.14}$$

or

$$\mu_{A_i^j}(x_i) = \exp\left\{-\left[\left(\frac{x_i - \mu_i^j}{\sigma^j}\right)^2\right]^{b_i^j}\right\} \tag{7.15}$$

where $[m_i^j, \sigma_i^j, b_i^j]$ is the parameter set to be tuned. In fact, continuous and piecewise differentiable functions, such as commonly used triangular or trapezoidal MFs, are also qualified candidates for node functions in this layer. Parameters in this layer are referred to as precondition parameters.

3. Layer 3: Every node in this layer is labeled Π, and multiplies the incoming
   signals $\mu_j = \mu_{A_1^j}(x_1) + \mu_{A_2^j}(x_2)$ and sends out the product. Each node output
   represents the firing strength of a rule.
4. Layer 4: Every node in this layer is labeled N and calculates the normalized
   firing strength of a rule. That is, the j-th node calculates the ratio of the j-th
   rule's firing strength of all the rules' firing strengths as:

$$\bar{\omega}_j = \mu_j \left/ \left(\mu_{A_1^j}(x_1) + \mu_{A_2^j}(x_2) \right) \right. \tag{7.16}$$

5. Layer 5: Every node j in this layer calculates the weighted consequent
   value as:

$$\bar{\omega}_j \left(a_0^j + a_1^j x_1 + a_2^j x_2 \right) \tag{7.17}$$

where $\bar{\omega}_j$ is the output of layer 4 and $\{a_0^j, a_1^j, a_2^j\}$ is the set to be tuned.
Parameters in this layer are referred to as consequent parameters.
6. Layer 6: The only node in this layer is labeled Σ, and it sums all incoming
   signals to obtain the final inferred result for the whole system (Lin and Lee,
   1996).

7.10.1   ANFIS PITFALLS

In many scientific papers, as many as 85 percent to 90 percent, it is not scientific
reasoning by the authors that is the core of the ANFIS methodology, but rather the
direct of software such as MAT LAB that yields results obtained according to
a variety of error definitions, such as mean absolute error (MAE), square-root error
(SRE), and similar criteria. Even the errors remain within acceptable limits but the
rational and logical basis of the application is erroneous. Unfortunately, in numerical
calculations, most of the time the error limit is considered as if they are the panacea
for any situation, which is a gross bias toward the philosophical and logical basis
of scientific research. In the following, several reflections are given from the open
literature to show the mechanical application of various methodologies, especially in
the domain of expert systems such as ANFIS. It is possible to increase the number of
such publications dramatically.

For instance, in Figure 7.29, MFs assigned for the input parameters are not logical from either a linguistic or numerical point of view. These are mechanically provided by ANFIS software, and many authors have accepted their validity without
criticism as to whether they are logically valid. The MFs have illogical sequence
and, unfortunately, the very basis of the universe of discourse division to MFs is not
taken into consideration. The MFs in Figure 7.29 have defaults, such as the leftmost
MFs—namely, "VC" and "C" for the input variable "distance," "D" have decreasing limbs toward "very low" "D" values. Here, the same illogical interpretation is
valid as in Figure 6.1 in Chapter 6. In addition, the rightmost MFs—namely, "VF"
for the same input variable and overlapping "H" and "VH" for the "charge weight

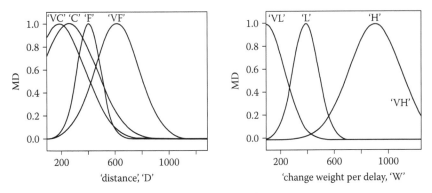

FIGURE 7.29 Membership functions assigned for the input parameter.

per delay," "W" variable has decreasing limbs toward "high" values, contrary to basic FL principles as explained in Chapter 3.

On the other hand, "VH" and "H" MFs for the input W" are almost equivalent within, say, ±2 percent error limits. This implies that the same set of data portion is attached with two different fuzzy sets with almost the same MDs, which cannot be acceptable in terms of FL principles.

In any modeling study, the question is: Is it enough to obtain numerical matching between the observed data and model results without considering the logical basis, or is it better to have concern for the logical validity at every stage of modeling and accordingly obtain a numerical match with the data? In many applications, the focus is on the former part, whereas the second part is ignored and hence the results and conclusions may appear as a mechanical fiddlings. The view that should be taken and emphasized in this book is the latter part of the question. For instance, in Figure 7.30a, the first MFs allocation should have the first and the last MFs with MDs equal to 1, which is not the case. The first and last MFs should have triangular shapes (Figure 7.30a) or trapezium shapes as in Figure 7.30b.

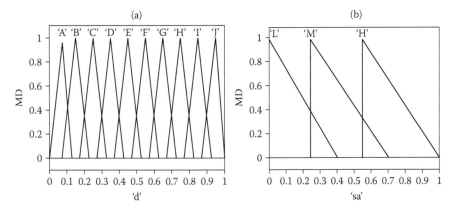

FIGURE 7.30 Fuzzy variables: (a) input, (b) output.

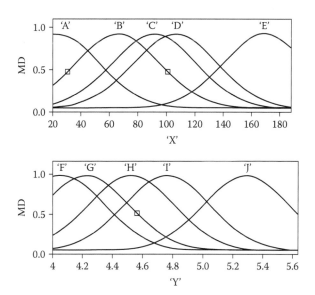

FIGURE 7.31 MFs of "*X*" and "*Y*."

In Figure 7.31, MFs for input "*X*" and output "*Y*" linguistic variables suffer from similar mistakes. Some of the MFs are nearly equivalent and especially the rightmost MFs—namely, "E" and "J" should attain MD 1 on the right limit.

REFERENCES

Jang, J.S.R. 1992. Self-learning fuzzy controller based on temporal back-propagation. *IEEE Trans. Neural Networks* 3(5): 714–723.

Lee, C.C. 1990. Fuzzy logic in control systems. Fuzzy logic controller, Parts I and II. *IEEE Trans. Syst. Man Cybern.* 20(20): 404–435.

Lin, C.-L. and Lee, C.S.G. 1996. *Neural Fuzzy Systems. A Neuro-Fuzzy Synergism to Intelligent Systems.* Prentice-Hall, Englewood Cliffs, NJ, 797 pp.

Mamdani, E.H. 1974. Application of fuzzy algorithms for simple dynamic plant. *Proc. IEE* 121: 1585–1588.

Mamdani, E.H. and Assilian, S. 1975. An experiment in linguistic synthesis with a fuzzy logic controller. *Int. J. Man Mach. Studies* 7(1): 1–13.

Sugeno, M., Ed. 1985. *Industrial Applications of Fuzzy Control,* North-Holland, New York.

Şen, Z. 1998. Fuzzy algorithm for estimation of solar irradiation from sunshine duration. *Solar Energy* 63(1): 39-49.

Şen, Z. 2004a. Genetik Algoritmalar ve Eniyileme Yöntemleri. (Genetic Algorithms and Optimization Methods). Su Vakfı Yayinlari (Turkish Water Foundation Publication), 235 pp. (in Turkish).

Şen, Z. 2004b. Yapay Sinir Ağları İlkeleriç (Artificial Neural Network Principles). Su Vakfı yayınları Yayınları (Turkish Water Foundation Publication), 285 pp. (in Turkish).

Takagi, T. and Sugeno, M. 1985. Fuzzy identification of systems and its applications to modeling and control. *IEEE Trans. on Systems, Man, and Cybern.* 15(1): 116–132.

Tsukamoto, Y. 1979. An approach to fuzzy reasoning method. In Gupta, M.M., Ragade, R.R., and Yager, R.R., Eds., *Advances in Fuzzy Set Theory and Applications*, North-Holland, Amsterdam, pp. 137–149.

PROBLEMS

7.1 Interpret the following composite consequent part of fuzzy inference system according to the following purposes:

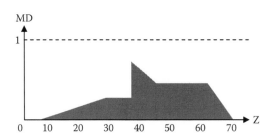

  (a) Flood protection
  (b) Drought mitigation
  (c) Most possible design
  (d) Without any prior information about the purpose

7.2 Draw the consequent part fuzzy inference from the following rule base:

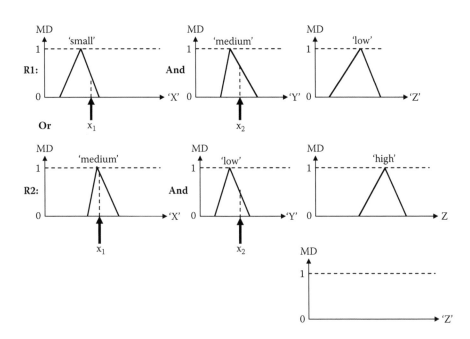

7.3 Write the Sugeno fuzzy inference system rules for the completion of the curve through the following five points:

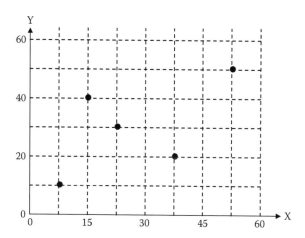

7.4 The input rule base of a hydrological event is given as follows:

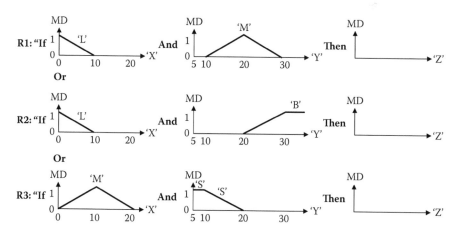

The output linguistic variable "Z" has three membership functions as in the following shape:

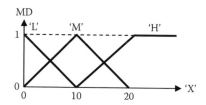

The data is also given in the following table:

Data Number	Input		Output
	"X"	"Y"	"Z"
1	7.1	8.2	5.3
2	3.2	26.5	6.9
3	1.5	31.4	11.5
4	19.2	5.5	3.2
5	8.3	2.9	8.3
6	12.6	7.4	5.7
7	19.3	11.5	13.9

Find the most suitable consequent membership function for each rule.

7.5 Suggest a fuzzy inference system (FIS) with three rules in the complete rule base that results in three trapeziums in the composite inference fuzzy set such that at least one of the consequent fuzzy sets is intact with other two.

8 Fuzzy Modeling of Hydrological Cycle Elements

8.1 GENERAL

Fuzzy logic, system, and modeling techniques were presented in previous chapters and now it is time to present some applications concerning various elements of the hydrological cycle. Each element is presented individually and then related to the closest hydrological element as a dependent variable. Almost all deterministic, stochastic equations, algorithms, and models can be fuzzified by considering reasonable fuzzification of the input and output linguistic variables. Additionally, their interrelationships can be put into a set of fuzzy rules as a rule base

8.2 SIMPLE EVAPORATION MODELING

The main hydrologic driving force for evaporation is the "temperature" variable, which can be divided into a range of linguistic categories such as "cold," "cool," "medium," "warm," and "hot" fuzzy classes, as already explained in Chapter 2 (Figure 8.1). Defining the bounds of these states is a bit subjective. An arbitrary crisp threshold might be set to separate "warm" from "hot" but this would result in a discontinuous change when the input value passes over that threshold. The way around this is to make the states fuzzy, that is, allow them to change gradually from one state to the next. For any given temperature value, almost always two successive MFs are triggered with different MDs. For instance, as shown in Figure 8.1, 0.2 "cool" and 0.8 "cold" MDs appear for T_{12} or 0.3 "warm" and 0.7 "hot" for T_{45}.

As the temperature changes, evaporation loses value in one MF while gaining value in the next. The output variable, "evaporation," is also defined by a fuzzy set that can have values like "v. low," "low," "medium," high," and "v. high," as in Figure 8.2.

The problem is a single-input-single-output modeling and hence, with given input and output variables, an action can be taken based on a set of rules that relate temperature variation to evaporation changes in the Mamdani FIS sense, such as:

R1: IF temperature is "hot" THEN evaporation is "v. high"

Generally, systems have many rules. Herein, R1 uses the MD of the temperature input in "hot" MF (see Chapter 6) to generate a result in the corresponding fuzzy set for the "evaporation" output, which is "high" MF. The individual rule consequent part result is used with the results of other rules to generate first a combined fuzzy

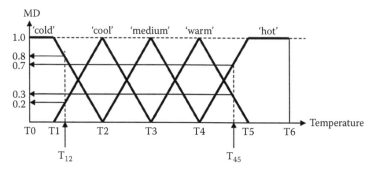

FIGURE 8.1 Temperature MFs.

set and then the crisp output value through a suitable defuzzification procedure (see Chapter 7). Obviously, the greater the MD of "hot," the higher the MD of "high," although this does not necessarily mean that the overall output from the valid rule base will be set to "high," because this is only one rule among many additional possible Mamdani FIS rules, including:

R1: "IF temperature is 'cold' THEN evaporation is 'v. low.'"

R2: "IF temperature is 'cool' THEN evaporation is 'low.'"

R3: "IF temperature is 'moderate' THEN evaporation is 'moderate.'"

R4: "IF temperature is 'warm' THEN evaporation is 'high.'"

R5: "IF temperature is 'hot' THEN evaporation is 'v. high.'"

In Figure 8.3, only R2, R3, and R5 are shown as a result of their triggerings with input data, say T_i. Each rule provides a result as a truth value of a particular MF for the output variable. In summary "min-max" FIS is used in finding the final aggregated overlap of all the three outcomes from each rule. The final evaporation result corresponding to T_i can be calculated by one of the defuzzification methods (Chapter 7).

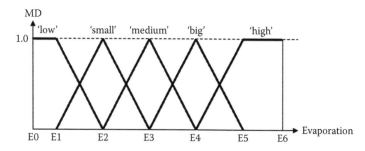

FIGURE 8.2 Evaporation MFs.

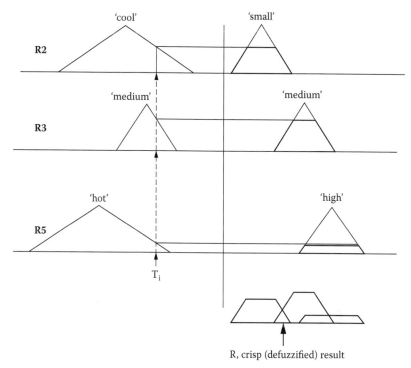

FIGURE 8.3 Evaporation FIS.

8.2.1 EVAPORATION ESTIMATION BY FIS

Evaluation of time series measurements is always associated with uncertainties due to data scarcity. Traditional methods rapidly become useless with decreasing data availability. A method based on fuzzy set theory and FIS can be applied to obtain more reliable information about the system from such scarce databases. To account for any uncertainty ingredient in the evaporation measurements, the FL and FIS concepts with a convenient rule base can be used for modeling and estimation purposes, respectively. The application of the methodology is presented for meteorological data recorded at the Eğirdir Lake location in a western province of Turkey, (Keskin et al., 2004). Lake Eğirdir (lat. 37° 60'–38° 43' N, long. 30° 30'–31° 37' E) is a freshwater lake located in Lakes District of Turkey, which is the second largest lake in the country with a surface area of 47,000 hm² and volume of 4.360×10^9 dm³ (Figure 8.4). It is a source of drinking water and irrigation activities in the region.

The necessary meteorological data to develop the fuzzy model approach is obtained from the Automated GroWeather Meteorological Station set up near Lake Eğirdir. Meteorological parameters include air and water temperatures, relative humidity, solar radiation, wind speed, air pressure, and sunshine hours, which are logged every 2 hours and integrated subsequently to obtain daily data.

The possible pairwise relationships between the input meteorological variables and output evaporation rates are investigated by inspecting scatter diagrams.

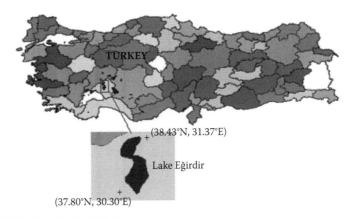

FIGURE 8.4 Map of Lake Eğirdir.

Accordingly, the effective variables on pan evaporation are arranged in order of their significance as air temperature T_a, water temperature T_w, solar radiation R_c, air pressure P_a, sunshine hours H_s, relative humidity R_h, and wind speed W_s, according to the degree of effectiveness on the evaporation measurements. First, a Fuzzy Evaporation Model (FEM) with four inputs is adopted, as shown in Figure 8.5. The adequacy of the FEM is evaluated by considering the coefficient of determination (R^2) based on the evaporation estimation errors as:

$$R^2 = \frac{e_o - e}{e_o} \tag{8.1}$$

where

$$e_o = \sum_{i=1}^{n} (E_{i(Pan)} - E_{mean})^2 \tag{8.2}$$

$$e = \sum_{i=1}^{n} (E_{i(Pan)} - E_{i(model)})^2 \tag{8.3}$$

Here, E_{mean} is the mean daily pan evaporation; $E_{i(Pan)}$ and $E_{i(model)}$ are daily pan measurement and FEM evaporation estimation, respectively; n is the number of data.

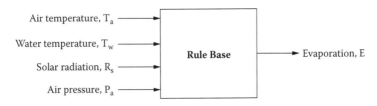

FIGURE 8.5 FEM configuration.

Root mean squared error (RMSE) values are also used as the index to check the ability of the model:

$$RMSE = \sqrt{\frac{1}{n}\sum_{i=1}^{n}(E_{i(Pan)} - E_{i(model)})^2}$$ (8.4)

Air temperature, water temperature, solar radiation, and evaporation variation domains are divided into eight, while air pressure into four triangular MFs (Figure 8.6). All inputs and output fuzzy sets are given in increasing magnitude with labels denoted by $T_a(i)$, $T_w(i)$, $R_c(i)$, and $E(i)$, where $i = 1, 2, \ldots, 8$ and $P_a(j)$ where $j = 1, 2, 3, 4$.

Historical data and expert knowledge are used to generate a rule base for the FEM. Rules are easy to define for extreme conditions, regardless of actual occurrences, because of the physical nature of the relationships. For intermediate MFs,

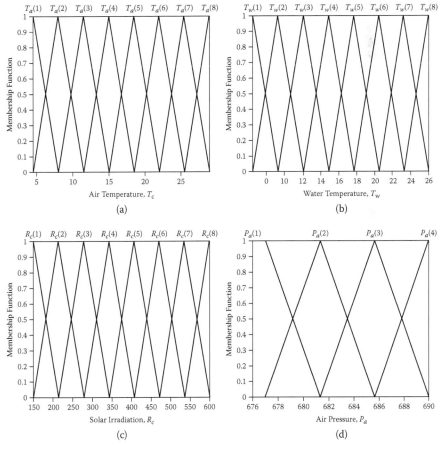

FIGURE 8.6 MFs: (a) air temperature, (b) water temperature, (c) solar radiation, and (d) air pressure.

rules are completed with the available data (Chapter 6). Although there are $8 \times 8 \times 8 \times 4 = 2,048$ possible antecedent combinations of rule bases, expert views and data usage have reduced the total rule base to 184 by combining antecedent input variables' fuzzy subsets with consequent evaporation fuzzy MFs.

Fuzzy clustering is also used to establish the rule base relationship between the input and output variables. The general form of each rule base is as follows:

$$\text{"IF } T_a \text{ is } T_a(i) \text{ AND } T_w \text{ is } T_w(i) \text{ AND } R_c \text{ is } R_c(i)$$
$$\text{AND } P_a \text{ is } P_a(j) \text{ THEN } E \text{ is } E(i)\text{."} \qquad (8.5)$$

The variable MFs in the antecedent part are combined into rules using the concept of "AND" ("ANDing"), which implies an FIS operator of "minimization" (Chapter 3).

The Sugeno FIS is applied at two stages, namely, training and testing (Sugeno, 1977; Chapter 7). During the training stage, 80 percent of the available data is used, whereas the remaining data set is employed for the verification (testing) stage.

Prior to execution of the model, normalization of the data X_i ($i = 1, 2, \ldots, n$) is achieved according to the following expression such that all data values fall between 0 and 1:

$$x_i = \frac{(X_i - X_{\min})}{(X_{\max} - X_{\min})} \qquad (8.6)$$

where x_i is the standardized, and $X_{\max}$ and $X_{\min}$ are the maximum and minimum measurement values, respectively. Such normalization procedures also render the data in dimensionless form. Furthermore, normalization also removes the arbitrary effects of similarity between objects or variables. MF types for inputs are selected as Gaussian bell whereas output is selected as a linear combination of input variables. Comparing the estimation of the FEM model with the measured pan evaporation values as in Figure 8.7a indicates that model performance is acceptable. Coefficients of determination (R^2) from Equation (8.1) and RMSE from Equation (8.4) are

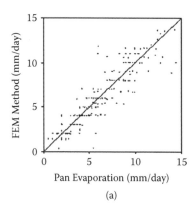

(a)

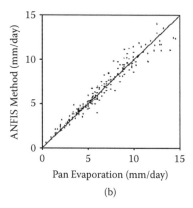

(b)

FIGURE 8.7 Scatter diagrams between daily pan evaporations: (a) FEM and (b) ANFIS.

calculated as 0.83 and 1.31 for FEM, respectively. The same set of input and output data has been subjected to automatic modeling through the ANFIS procedure using readily available MATLAB software; the results are shown in Figure 8.7b, where the R^2 and RMSE values are 0.96 and 0.65, respectively. It can be seen from Figure 8.7 that both methods yield scatterplots that are close to a 45° straight line, implying that there are no model bias effects.

8.3   INFILTRATION RATE MODEL

Logical crisp relationships are well known in hydrology literature between the infiltration rate f and the rainfall intensity i (after each storm rainfall) and fall in three categories:

1. $f > i$ implies that there is infiltration without runoff.
2. $f = i$ means that there is infiltration rate without runoff but the soil is just saturated.
3. $f < i$ shows that there is no infiltration because the soil is over-saturated.

These implications help when considering the ratio, $\alpha = i/f$, which implies an infiltration phenomenon only in cases where $0 < \alpha < 1$; otherwise for $\alpha \geq 1$, only runoff occurs, that is, the soil is completely saturated. This implies at any soil point a crisp step function or crisp threshold for the α parameter, as in Figure 8.8a. The infiltration capacity of a soil at a point is independent of rainfall intensity. This CL point solution cannot be applied in practice to the land surrounding the point due to the soil heterogeneity and unisotropy. How many point solutions are needed within a watershed to represent the aggregate behavior of watershed lands is an important question.

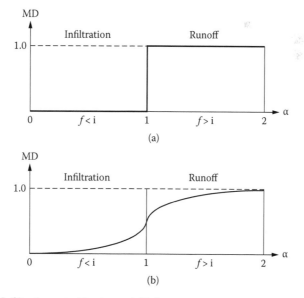

FIGURE 8.8 Infiltration sets: (a) crisp and (b) fuzzy.

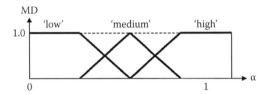

FIGURE 8.9 Fuzzy infiltration sets.

Soil structure, together with areal variability in topography and vegetation, limit the direct use of infiltration functions to areas beyond a point and this is why CL thresholds as in Figure 8.8a are a problem. Soil is not a really homogeneous and isotropic porous medium, but has been subjected to various physical, chemical, and biological processes that might have developed structure in the larger voids; these processes allow flowing water to bypass parts of the soil matrix. Field measurements of infiltration rates often show large variations, even within small areas. This implies that fuzziness is embedded in each infiltration test result and hence the shape in Figure 8.8a may remain almost the same but with right or left shifts of the threshold value in α. In actual situations, there should be a smooth transition (threshold) around the condition $\alpha = 1$, and a plausible case is presented in Figure 8.8b, which implies that the transition from infiltration to runoff is not a crisp event. To describe the fuzzy properties of the infiltration rate, Figure 8.9 presents fuzzy partitions in terms of "low," "medium," and "high" for $\alpha < 1$.

Infiltration tests in the field always include imprecision and expert views of hydrologists help in reaching reliable evaluations and estimations. The imprecision includes various effects, such as the test performance, instrument suitability, soil and subsoil heterogeneities, and water-addition into the inner cylinder during a double-infiltrometer test in the field. The available empirical and analytical approaches consider the soil structure as homogeneous and isotropic. For such ideal cases, Horton (1940) developed an empirical relationship (based on CL, of course) describing the dependence of the infiltration rate f on time t:

$$f = f_c + (f_0 - f_c)e^{-kt} \tag{8.7}$$

in which f_c and f_0 are the final and the initial (infiltration capacity) infiltration rates, respectively. The exponential constant k gives the decay rate with time depending on soil properties. In practice, difficulties in the application of this formula lie in the accurate estimation of the parameters f_0, f_c, and k.

Methods for dealing with surficial hydrologic processes such as the infiltration process need to:

1. Solve for key processes such as infiltration rate at many points on the watershed surface. When different permeabilities, moisture storages, etc. are used at each point on the watershed surface, point-to-point variability in processes should generate continuously variable and physically realistic aggregate watershed behavior.

TABLE 8.1

Infiltration Rate Data

Time (min)	Infiltration Rate (cm/min)
5	4.71
10	3.77
15	2.12
20	1.80
25	1.41
30	1.35
40	1.29
50	1.06
60	0.82
90	0.78
120	0.66
150	0.58

2. Use fuzzy sets or continuous functions to interrelate variables in hydrologic process calculations and model aggregate behavior directly. If the soil properties at a point are defined, a numerical solution for the infiltration rate at this point can be calculated according to Equation (8.7) in a crisp manner.

In an alluvium area of arid regions, the infiltration rate calculations are obtained after the field test and the results are represented in Table 8.1, with its graphical representation in Figure 8.10 showing Horton equation trend.

Figure 8.10 indicates that although there is a generally decreasing trend in the infiltration rate over time, there are also local deviations. In classical hydrological

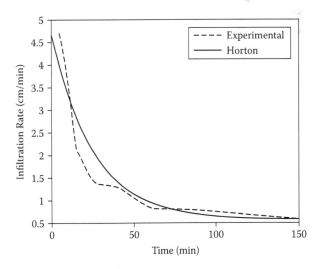

FIGURE 8.10 Experimental infiltration rate patterns.

calculations, the fitting of Equation (8.7) to the field data smoothes out these local variations, and yields a smooth curve with $f_0 = 4.71$ cm/min, $f_c = 0.58$ cm/min, and $k = 0.04$.

It is suggested herein to process the experimental infiltration data using FL methods; these methods do not have parameters such as f_0, f_c, and k but instead have a set of rules that embeds relationships between infiltration rate and time. In general, the relationship implies that as the time increases, the infiltration rate decreases. This linguistic information is global and non-fuzzy (crisp). Its fuzzification is necessary for the depiction of local deviations. The fuzzifications of time and the infiltration rate variables can be achieved by linguistic terms such as *early*, *moderate*, and *late* time periods corresponding to *high*, *medium*, and *low* infiltration rates. These linguistic words help in writing the following fuzzy rule base similar to Mamdani FIS:

R1: "IF time is 'early' THEN infiltration rate is 'high.'"

R2: "IF time is 'moderate' THEN infiltration rate is 'medium.'"

R3: "IF time is 'late' THEN infiltration rate is 'low.'"

These are the only logical rules out of a total $3 \times 3 = 9$ rules that relate infiltration process variation to time. For instance, the following rule is one of the illogical (invalid) rules for the infiltration rate change with time.

R4: "If time is 'early' THEN infiltration rate is 'high.'"

The logical infiltration rate time variation rules are presented in Figure 8.11 with overlapping rectangular areas based on the triangular MFs similar to what was explained in Chapter 5 concerning inversely proportionality (Section 5.6.2 in Chapter 5).

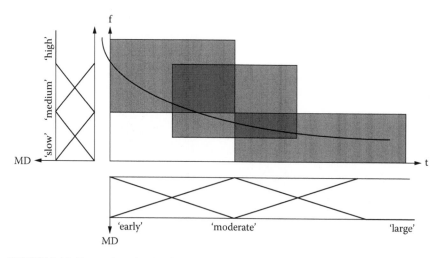

FIGURE 8.11 Fuzzy domain of infiltration rate.

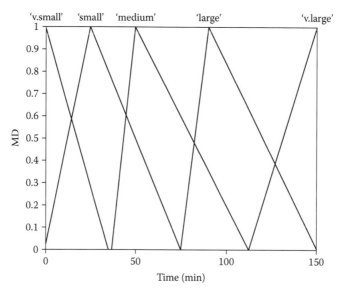

FIGURE 8.12 Sugeno constant consequent FIS MFs.

It is possible to model the infiltration rates according to Sugeno FIS by considering the consequent parts of the rule bases as either constant or linear functions of the input variable, which is time t. First, the constant consequent part Sugeno method is applied (see Chapter 7). For this purpose, the time domain is divided into five triangular MFs as "very small," "small," "medium," "large," and "very large" time linguistic fuzzy categories as shown in Figure 8.12.

Figure 8.12 includes the trained MFs with the resultant infiltration rate data from Table 8.1 through the following fuzzy rule bases:

R1: "IF time, t, is 'very small' THEN $f = 4.71$."

R2: "IF time, t, is 'small' THEN $f = 1.30$."

R3: "IF time, t, is 'medium' THEN $f = 0.80$."

R4: "IF time, t, is 'large' THEN $f = 0.65$."

R5: "IF time, t, is 'very large' THEN $f = 0.58$."

After the application of the Sugeno constant consequent FIS, the infiltration curve is shown in Figure 8.13 by a solid line together with the Horton analytical model solution as a broken line. It is obvious that the two curves follow each other rather closely within less than 10 percent error, especially at "medium" and "small" fuzzy time domains. The same data is subjected to linear consequent Sugeno FIS, which requires linear line succession of the general infiltration rate curve as in Figure 8.14. It is therefore necessary to establish some approximate

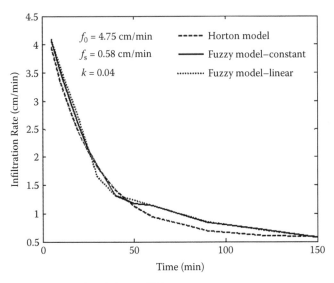

FIGURE 8.13 Horton model and Sugeno FIRs.

linear portions to the overall infiltration trend, and three of them are shown in the same figure. To model the infiltration rate through linear consequent Sugeno FIS, four triangular MFs are adopted and their final forms, after the necessary trial-and-error effort for finding the best match to the data in Table 8.1, are presented in Figure 8.15.

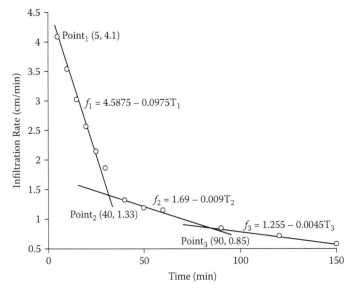

FIGURE 8.14 Linear consequent Sugeno FIS preparation lines.

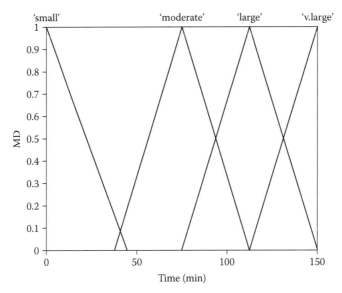

FIGURE 8.15 Sugeno constant consequent FIS MFs.

The fuzzy rule base of this model is presented as an inverse relationship of four statements:

R1: "IF time, t, is small THEN infiltration rate is $f = -0.0975t + 4.588$."

R2: "IF time, t, is moderate THEN infiltration rate is $f = -0.0090t + 1.690$."

R3: "IF time, t, is large THEN infiltration rate is $f = -0.0045t + 1.255$."

R4: "IF time, t, is small THEN infiltration rate is $f = -0.0045t + 1.255$."

Final infiltration rate variation with time according to the linear consequent Sugeno model is given in Figure 8.13, which has better closeness to the Horton model than the Sugeno constant consequent model.

8.4 RAINFALL AMOUNT PREDICTION

Rainfall estimation has been achieved in hydrology for several decades by stochastic processes. In this estimation process, there is a set of assumptions such as stationary, ergodicity, normality, linearity, etc. It is possible to make rainfall estimation by an FL algorithm without such assumptions. Herein, an FIS model algorithm is presented for monthly rainfall estimations.

The algorithm estimates the rainfall at, for example, month $k + 1$ from the rainfall amount at month k. As a simple linguistic model, it is possible to state that the rainfall amount $r(k + 1)$ is equal to the rainfall amount $r(k)$ plus the rainfall change $\Delta r(k)$, which may be either "positive" or not.

$$r(k + 1) = r(k) + \Delta r(k) \tag{8.8}$$

The rainfall change can be expressed as the summation of the rainfall amount at time $(k + 1)$ with subtraction (or addition) of a random variable $\varepsilon(k)$:

$$\Delta r(k + 1) = r(k) + \varepsilon(k) \tag{8.9}$$

These two equations can be written in matrix form as:

$$\begin{bmatrix} r \\ \Delta r \end{bmatrix}_{k+1} = \begin{bmatrix} 1 & 1 \\ 1 & 0 \end{bmatrix} \begin{bmatrix} r \\ \Delta r \end{bmatrix}_k + \begin{bmatrix} 0 \\ \varepsilon \end{bmatrix}_k \tag{8.10}$$

In this equation system, all the variables are taken as standardized values, which means that after the subtraction of arithmetic average from the actual rainfall amounts, the difference is divided by the standard deviation. The standardized variables of rainfall assume negative and positive values around zero.

To solve the problem, it is necessary to decide about the variation domain of each variable, that is, the universe of discourse. The variability domains are adopted as -2 cm $< r(k) < +2$ cm and $-5 < \Delta r(k) < +5$.

The following steps expose the FL solution to the rainfall amount estimation problem:

1. Rainfall is considered as having three MFs with "positive" ("P"), "zero" ("Z"), and "negative" ("N") linguistic labels and the triangular MF forms as in Figure 8.16. The rainfall increment variable is also given in terms of another three MFs (Figure 8.17). As shown in Figure 8.18, the error variable discourse domain from -24 to $+24$ is divided into seven MFs ("negative very big" ['NVB'], "negative big" ['NB'], "negative" ['N'], "zero" ['Z'], "positive" ['P'], "positive big" ['PB'], and "positive very big" ['PVB']) will be effective in the dynamic procedure.
2. The entire FAM (rule base) is given in Table 8.2 for 3×3 input variable joint MF regions. This table indicates the decision surface of the prediction problem based on the decision variables, which are $r(k)$ and $\Delta r(k)$,

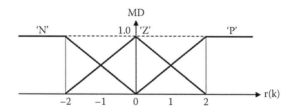

FIGURE 8.16 Standard rainfall MF.

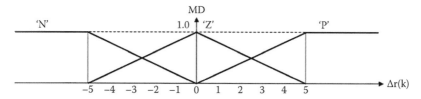

FIGURE 8.17 Rainfall increment MFs.

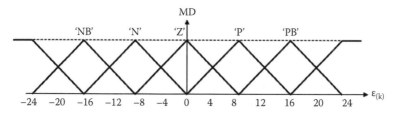

FIGURE 8.18 Random error MFs.

3. Now everything is ready for the activation of this prediction problem explicitly. First, the following initial conditions are considered: $r(0) = 1$ cm and $\Delta r(0) = -4$ mm. The procedure will be achieved for three steps, namely, $k = 0, 1, 2$. Each run will result in an MF based on the two decision variable values, $r(k)$ and $\Delta r(k)$. The FAM in Table 8.2 will produce the fuzzy consequence of each run in terms of the $\varepsilon(k)$ fuzzy subset. The resulting fuzzy output will be defuzzified by the centroid method (see Chapter 7) and then the recursive equations will be used for the next decision variable numerical values.

Figures 8.19 and 8.20 indicate triggering of the given initial data on the input variable MFs. It is possible to write the following rules from the FAM:

R1: "IF $r(k)$ is 'P' AND $\Delta r(k)$ is 'Z' THEN $r(k+1)$ is 'P.'"

R2: "IF $r(k)$ is 'P' AND $\Delta r(k)$ is 'N' THEN $r(k+1)$ is 'Z.'"

TABLE 8.2

Fuzzy Associative Memory (FAM)

$r(k)$ \ $\Delta r(k)$	"P"	"Z"	"N"
"P"	"PB"	"P"	"Z"
"Z"	"P"	"Z"	"N"
"N"	"Z"	"N"	"NB"

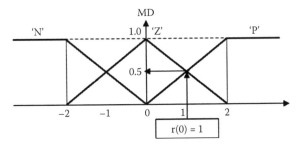

FIGURE 8.19 Rainfall MFs.

R3: "IF $r(k)$ is 'Z' AND $\Delta r(k)$ is 'Z' THEN $r(k+1)$ is 'Z.'"

R4: "IF $r(k)$ is 'Z' AND $\Delta r(k)$ is 'N' THEN $r(k+1)$ is 'N.'"

According to the triggering of initial values in Figures 8.19 and 8.20 and consideration of the minimization fuzzy operator procedure for the antecedent parts, the consequent parts result, respectively, for each rule:

$$\min(0.5, 0.2) = 0.2\text{"P"}$$

$$\min(0.5, 0.8) = 0.5\text{"Z"}$$

$$\min(0.5, 0.2) = 0.2\text{"Z"}$$

$$\min(0.5, 0.8) = 0.5\text{"N"}$$

Figure 8.21 indicates the union of the truncated consequent for the output control variable $r(k+1)$. Finally, Figure 8.22 presents the defuzzification value.

By taking into consideration the defuzzified value $\varepsilon(0) = -2$ together with Equations (8.8) and (8.9), one can find the first monthly rainfall estimation as:

$$r(1) = r(0) + \Delta r(0) = 1 - 4 = -3$$

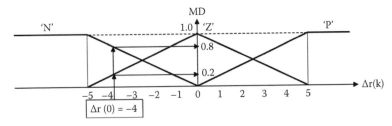

FIGURE 8.20 Rainfall increment MFs.

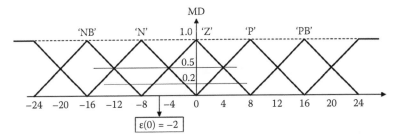

FIGURE 8.21 Random error MFs.

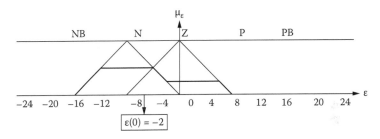

FIGURE 8.22 Union of consequents and defuzzification.

and

$$\Delta r(1) = r(1) - \varepsilon(0) = -3 - (-2) = -1$$

These estimates trigger the input variables as in Figures 8.23 and 8.24.

Again, consideration of the FAM relationships in Table 8.2 gives the following fuzzy-rule base:

R1: "IF $r(k)$ is 'N' AND $\Delta r(k)$ is 'N' THEN ε is 'NB.'"

R2: "IF $r(k)$ is 'N' AND $\Delta r(k)$ is 'Z' THEN ε is 'N.'"

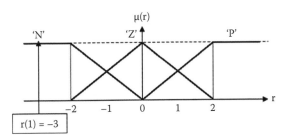

FIGURE 8.23 Rainfall MFs.

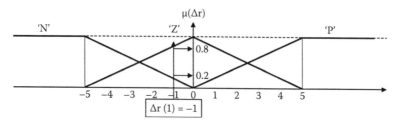

FIGURE 8.24 Angle change MF.

The minimization operation of the antecedent fuzzy subsets leads, respectively, for each rule, to:

$$\min(1, 0.2) = 0.2"NB"$$

$$\min(1, 0.8) = 0.8"N"$$

The union of these two truncated fuzzy subsets (see Figure 8.25) and their defuzzification yields $\varepsilon(1) = -9.6$.

Now the new estimates can be obtained from the recurrence of Equations (8.8) and (8.9) as:

$$r(2) = r(1) + \Delta r(1) = -3 - 1 = -4$$

and

$$\Delta r(2) = r(2) - \varepsilon(1) = -4 - (-9.6) = +5.6$$

Again, consideration of the FAM rules in Table 8.2 with these initial values leads to the following single valid rule –base:

R1: "IF $r(k)$ is 'N' AND $\Delta r(k)$ is 'P' THEN ε is 'Z.'"

with the minimization operator as $\min(1, 1) = 1$ "Z." Hence, the next estimation calculations lead to:

$$r(3) = r(2) + \Delta r(2) = -4 + 5.6 = 1.6$$

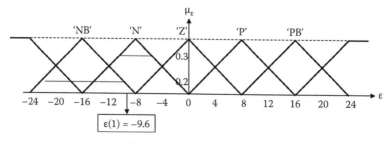

FIGURE 8.25 Random error MFs.

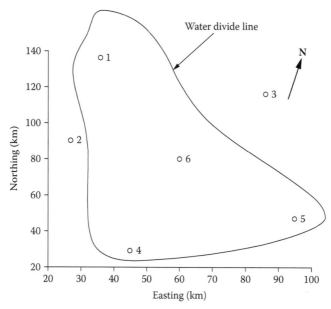

FIGURE 8.26 Location map.

and

$$\Delta r(2) = r(3) - \varepsilon(1) = -1.6 - (-9.6) = +8.0$$

The remaining steps are left to the reader for execution, and in this manner the monthly rainfall estimate can be continued until any time instant that he/she wishes.

8.4.1 Areal Rainfall Estimation

The application of fuzzy modeling for spatial rainfall estimation is presented for a hypothetical drainage area as shown in Figure 8.26 with easting and northing locations of six raingauge stations, where four of the stations are within the basin.

The annual rainfall R amounts at each raingauge point and corresponding location information (easting and northing distances, D_e and D_n) are given in Table 8.3.

TABLE 8.3
Raingauge Information

Raingauge Number (i)	D_e (km)	D_n (km)	R (cm)
1	36	136	63
2	27	90	172
3	86	116	97
4	45	29	200
5	95	47	118
6	60	80	55

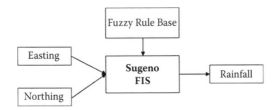

FIGURE 8.27 Sugeno spatial rainfall FIS.

The Sugeno method can be visualized as in Figure 8.27. The backbone of such a system is the fuzzy rule base, which relates the easting and northing fuzzy MFs to rainfall amount MFs. Herein, Sugeno FIS is used and for this reason, the rainfall is represented by the actual rainfall amounts recorded at each location, which implies that there are no MFs for the consequent part in the fuzzy rule base.

Because there are six stations, there will be six MFs for easting and northing directions. Although the MFs can be adopted as any normal fuzzy set, here Gaussian fuzzy sets (Equation 3.21) are preferred due to the storm rainfall occurrence shape. In practice, the scale parameter m with dispersion parameter s can be calculated according to the problem at hand. The dispersion parameter can be calculated as the standard deviation of a given precipitation series. Otherwise, it is chosen in an expert manner such that the overall areal rainfall estimation error becomes minimal.

Figure 8.28 and Figure 8.29 show adopted fuzzy MFs for easting and northing, respectively. In the easting direction, the six fuzzy sets are "short" ("S"), "medium short" ("MS"), "medium" ("M"), "medium large" ("ML"), "large" ("L"), and "very large" ("VL"). Similar assignments are valid for the northing distance variable. The

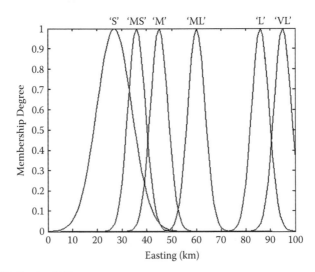

FIGURE 8.28 Easting MFs.

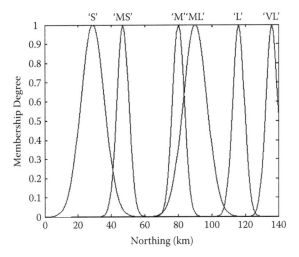

FIGURE 8.29 Northing MFs.

following points are considered in the construction of each MF:

1. Each fuzzy set has its peak with MD equal to 1 at the raingauge location.
2. The dispersion coefficient of each fuzzy set is adopted according to two criteria:
   a. If there are historical rainfall records at each station, then the standard deviation of the past rainfall records is adopted as the dispersion coefficient (s) of each MF.
   b. By training the fuzzy system so as to obtain the least sum of squares deviation from the observed rainfall values at measurement sites.
3. Deduction of the fuzzy rule base by considering the raingauge station configuration, rainfall records, and the meteorological facts about the environment.

The rule base can be written as follows for all the logical combinations of the easting and northing locations by considering the station location with the rainfall records as in Table 8.3 at each station according to Sugeno FIS:

R1: "IF Easting is 'S' AND Northing is 'ML' THEN rainfall is equal to 172."

R2: "IF Easting is 'MS' AND Northing is 'VL' THEN precipitation is equal to 63."

R3: "IF Easting is 'M' AND Northing is 'S' THEN rainfall is equal to 200."

R4: "IF Easting is 'ML' AND Northing is 'M' THEN rainfall is equal to 55."

R5: "IF Easting is 'L' AND Northing is 'L' THEN rainfall is equal to 97."

R6: "IF Easting is 'VL' AND Northing is 'MS' THEN rainfall is equal to 118."

The easting and northing fuzzy MFs are combined by the multiplication procedure through the general Sugeno FIS. The reason for adopting Sugeno FIS is due to the fact that the system outputs from Figure 8.27 appear as a weighted average (similar to Equation 7.2):

$$P(x_i, y_i) = \frac{\sum\limits_{i=1}^{N} r_i \prod\limits_{j=1}^{n} \left\{ \exp\left[-\left(\dfrac{x-x_i}{s_{xi}}\right)^2 \right] \exp\left[-\left(\dfrac{y-x_j}{s_{yj}}\right)^2 \right] \right\}}{\sum\limits_{i=1}^{N} \prod\limits_{j=1}^{n} \left\{ \exp\left[-\left(\dfrac{x-x_i}{s_{xi}}\right)^2 \right] \exp\left[-\left(\dfrac{y-x_j}{s_{yj}}\right)^2 \right] \right\}} \qquad (8.11)$$

where $P(x_i, y_i)$ is the spatial rainfall estimation at a point given its easting x_i and northing y distances. In this expression, the weightings are given as:

$$w_{i,j} = \sum\limits_{i=1}^{N} \prod\limits_{j=1}^{N} \left\{ \exp\left[-\left(\dfrac{x-x_i}{s_{xi}}\right)^2 \right] \exp\left[-\left(\dfrac{y-x_j}{s_{yj}}\right)^2 \right] \right\} \qquad (8.12)$$

Substitution of easting and northing values into this expression, together with the values from Table 8.3, yields the spatial rainfall estimation values. It is noteworthy that this fuzzy inference system and the spatial rainfall estimation procedure yield the exact rainfall amounts at existing raingauge stations. Table 8.4 includes the areal rainfall estimation amounts at a set of desired points.

TABLE 8.4
Spatial Rainfall Estimation with FIS

Station Number	Easting (km)	Northing (km)	Rainfall (cm)
(i)	(xᵢ)	(yᵢ)	(rᵢ)
1	36	136	63
2	27	90	172
3	86	116	97
4	45	29	200
5	95	47	118
6	60	80	55
7	52	60	55.1
8	50	120	63.2
9	80	60	118
10	70	140	97

8.5   RAINFALL–RUNOFF RELATIONSHIP

A common and simple approach in the assessment of rainfall–runoff relationships is the use of classical regression analysis, which is relied upon as a remedy for very complicated rainfall–runoff processes in the preliminary stages of runoff estimation for hydrologic design. However, prior to its use, the following imbedded hydrologic assumptions, comments, and simplifications should be considered:

1. In simple regression relationships, rainfall is assumed as uniformly distributed over the drainage area. Such an assumption might be valid for small areas, but as the area increases, the validity of this approach must be questioned (Şen and Wagdani, 2008). Consequently, more uncertainties become included in the overall rainfall–runoff transformation process,
2. Depending on the antecedent soil and surface conditions of the drainage area, the portion of the rainfall that appears as direct runoff will be different even when the peak rainfall amounts are the same. This indicates that the transformation to runoff is not static, but rather a dynamic process according to the environmental conditions. For instance, during wet periods, the conditions are different than during dry spells. It should be noted here that the words *wet* and *dry* are linguistically fuzzy in contents.
3. Logically, the amount of rainfall is greater than the generated runoff from the same storm, and consequently, the proportionality factor (runoff/rainfall), which is known as the runoff coefficient in hydrology, assumes values between 0 and 1. However, the runoff coefficient is not a constant throughout the year; it also depends on the antecedent conditions (Kadioglu and Şen, 2001). It is not possible to consider such variations in the coefficients through regression analysis where the calculations are based on the numerical data only without linguistic information. However, the FIS approach provides a basis for considering such information through vaguely defined MFs.

In addition to these hydrological requirements, most often given the measurements of rainfall and runoff, the determination of the rainfall-runoff relationship is a first-stage analysis, which requires a suitable methodology. Because the rainfall and runoff measurements show random fluctuations around averages, the convenient methodology should be based on uncertainty techniques such as the FL approaches. Most often, a simple linear relationship between the dependent runoff and independent rainfall variables is preferred. For any regression application to determine the rainfall–runoff relationship, the following steps are necessary in the calculations:

1. In practice, the rainfall and runoff measurements are plotted on a Cartesian coordinate system and the result is the scatter of points. In fact, each one of these points corresponds to different antecedent or environmental conditions. However, in regression line fitting, such a distinction is not considered, and each point in the scatter is treated equally, leading to the best straight line. Use of the regression approach brings into view additional restrictive assumptions, such as the equivalence of variances, independence

of deviations of each scatter point from the fitted regression line, Gaussian distribution, etc. If these procedural assumptions are not satisfied, then the regression approach leads to biased rainfall–runoff relationships.

2. In addition, the regression methodology yields crisp values for parameters. This implies that irrespective of seasonality as an important factor, its influence on vegetation cover and infiltration rates, the parameters are considered as having the same values for different seasons or months.

3. Although the scatter diagram shows the random behavior of the drainage basin, in the sense that the rainfall–runoff transformation does not change significantly with the physical characteristics of the basin, the regression parameters are expected to vary with the duration of rainfall and antecedent conditions, which are not evident in the regression model.

In summary, all these points include rather fuzzy uncertainties.

8.5.1 Crisp Rainfall–Runoff Relationship

In general, the linear regression is fitted to the scatter of rainfall–runoff points conventionally with constant model parameter values. However, such an approach ignores the dynamic behavior of the rainfall–runoff process, and consequently the variations in the data are rendered to a completely deterministic world without any linguistic information. It is well known that the rainfall–runoff process is dynamic and non-linear in nature, where proportionality and superposition principles do not apply (Kundzewicz and Napieorkowski, 1986). The connection of the rainfall–runoff points in the logical monthly sequence leads to irregular polygons on the coordinate systems (Kadıoğlu and Şen, 2001). Such a hypothetical rainfall and runoff scatter diagram with 12 points, and their successive corrections, are presented in Figure 8.30, which yield a set of linguistic information. Because the 12-sided shape is in the form of an irregular polygon, it is referred to also as the rainfall–runoff polygon.

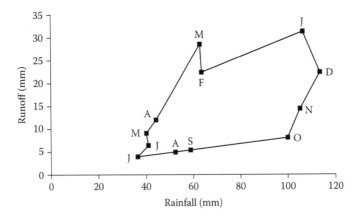

FIGURE 8.30 Monthly rainfall–runoff polygons.

The following linguistic information can be derived from such polygons in an FL manner with vague interpretations:

1. The lengths of polygon sides indicate the change in average values of rainfall or runoff for consecutive months in terms of "little," "significant," "moderate," and "big" changes.
2. The length of each polygon side indicates the value of the runoff coefficient between rainfall and runoff during consecutive months.
3. The closeness of the slope of each side to the vertical or horizontal indicates the relative proportions of the rainfall and runoff. Similar interpretations for all the months during 1 year provide a basis for qualitative interpretations about the rainfall–runoff occurrences in a catchment. In any regression approach, these differences in the proportionalities during one year are not taken into consideration. However, the FL approach accounts for such differences, and it is possible to classify the slopes as "very small," "small," "moderate," "high," and "very high" in terms of fuzzy subsets.
4. A long each side of the polygon, runoff is assumed to change linearly with rainfall. Such a linearity assumption during time intervals smaller than 1 year yields more reliable results in the runoff volume calculations. The polygon constitutes finite straight-line portions for the validity of a linearity assumption on a monthly basis. Practically, if all the sides fall along a single direction within ±5 percent or ±10 percent deviations, then the corners in the polygon diagram might be considered as scattered along a regression straight line, which represents the classical rainfall–runoff relationship. The narrower the polygon, the more representative the regression approach for rainfall–runoff modeling. On the contrary, wide polygons imply heterogeneous temporal variations, dynamism, and nonlinearity in rainfall–runoff relationships for the catchment"s area—and hence the applicability of FL modeling,
5. The smaller the area of the polygon, the more consistent the monthly rainfall and the more reliable the regression estimation of the resulting runoff. Otherwise, the results are not reliable and, instead, the fuzzy approach must be employed in finding the rainfall–runoff transformation relationship.

8.5.2 Fuzzy Rainfall–Runoff Relationship

Rainfall and runoff variables are considered in five partial subgroups: "low" ("L"), "medium low" ("ML"), "medium" ("M"), "medium high" ("MH"), and "high" ("H"). A small number of fuzzy sub-group selection leads to unrepresentative predictions whereas large numbers imply unnecessary calculations. In practical studies, most often the sub-group number is selected preliminarily as 4 or 5 and about 7 at the maximum (Şen, 2001). Five sub-groups in each variable imply that there are $5 \times 5 = 25$ different partial relationship pairs that may be considered between the rainfall and runoff variables. However, many such partial relationships are not physically plausible. For instance, if the excess rainfall is "high," it is not possible to state that the direct runoff is "low" or even "medium." Figure 8.31 shows the relative positions of the aforementioned fuzzy words.

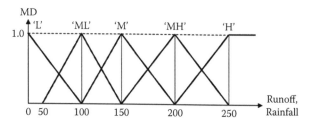

FIGURE 8.31 Hypothetical fuzzy sub-groups of rainfall and runoff.

Because the rainfall–runoff relationship, in general, has a direct proportionality feature, it is possible to write the following five rules for the description of fuzzy rainfall–runoff modeling:

R1: "IF rainfall is 'L' THEN runoff is 'L.'"

R2: "IF rainfall is 'ML' THEN runoff is 'ML.'"

R3: "IF rainfall is 'M' THEN runoff is 'M.'"

R4: "IF rainfall is 'MH' THEN runoff is 'MH.'"

R5: "IF rainfall is 'H' THEN runoff is 'H.'"

The general appearance of such a model is given in Figure 8.32 where rather than a regression model of the classical calculations, a fuzzy region of rainfall–runoff relationships is considered with uncertainty domain.

To show the application of FIS, two rules (R1 and R2) are shown in Figure 8.33 with rainfall intensity $i = 22.5$ mm triggering. The consequent part of each runoff variable appears as a truncated trapezium for each rule.

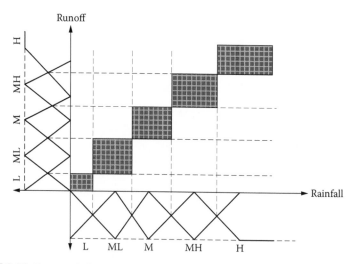

FIGURE 8.32 Fuzzy rule-based rainfall–runoff relationship domain.

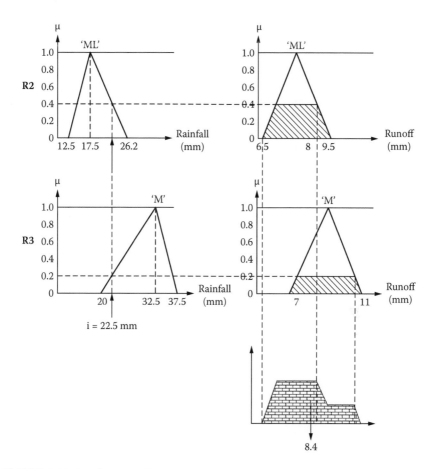

FIGURE 8.33 Rainfall–runoff rules.

The overlapping of these two truncated trapeziums indicates the combined inference from these two rules as in the lower part of the same figure, which is represented in Figure 8.34 where $A1$, $A2$, $A3$, $A4$, $A5$, and $A6$ indicate triangular and rectangular sub-areas in the composite fuzzy inference set. For hydrologic design purposes, it is necessary to deduce a single value, which is referred to as defuzzification in fuzzy systems terminology (see Chapter 7). Although there are various defuzzification methods, the most common method is that of centroid defuzzification (Ross, 1995; S¸en, 2001). By applying the centroid defuzzification formula to the fuzzy inference set in Figure 8.34, it is possible to obtain defuzzification value as:

$$\hat{x} = \frac{\sum_{i=1}^{6} x_i A_i}{\sum_{i=1}^{6} A_i} = \frac{0.05*6.67+0.90*7.88+0.05*9.02+0.10*9.25+0.25*10.13+0.03*10.83}{0.05+0.90+0.05+0.10+0.25+0.03}$$

$$= 8.4$$

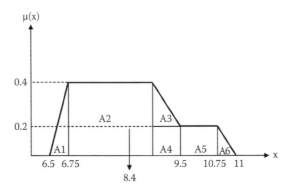

FIGURE 8.34 Fuzzy inference subset.

Another application of CL and FL methodologies is presented for two different drainage basins, namely, Terkos on the European and Omerli on the Asian side of Istanbul (Figure 8.35). For better water supply to Istanbul, these catchments are connected jointly through a submarine pipe between the two continents. Measurements in both catchments for almost 20 years help to depict rather fuzzy rainfall–runoff relationships.

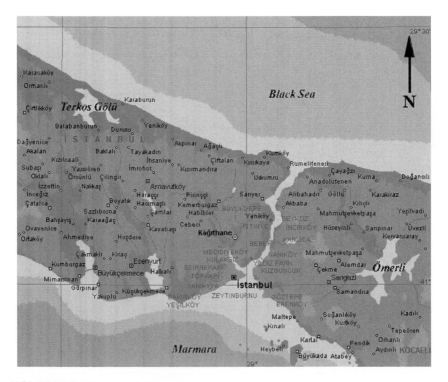

FIGURE 8.35 Catchment location map.

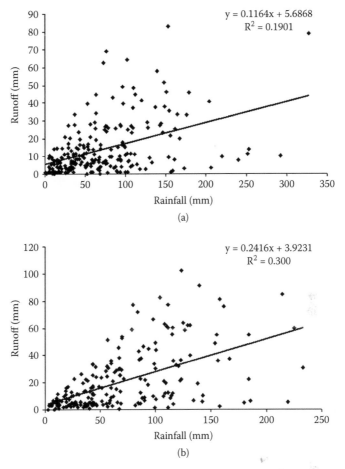

FIGURE 8.36 Classical rainfall–runoff relations: (a) Terkos and (b) Ömerli catchments.

Figure 8.36 shows the classical regression lines fittings for the determination of the runoff coefficient in each catchment. It is obvious that the scatter of rainfall–runoff data is considerable, which suggests instability in the rainfall–runoff relationship. In addition, the prerequisite assumptions in any classical regression procedure, especially the variance constancy (homoscedasticity), are not met because in each scatter diagram, the runoff variance is small for small rainfall values, it increases as the runoff increases. Hence, the classical regression lines cannot provide reliable relationships. Table 8.5 and Table 8.6 give the monthly averages and standard deviations, respectively, of rainfall and runoff for each catchment. The application of the Mamdani FIS model with centroid defuzzification method is performed for each catchment through written software and the results are presented in Tables 8.7 for the Terkos catchment. The relative errors for each runoff prediction through the classical regression and fuzzy models are also presented. It is observed that, invariably, the fuzzy approach provides better estimates from the classical regression

TABLE 8.5
Monthly Rainfall Data Statistics

| Month | Catchments | | | |
| | Terkos | | Ömerli | |
	Ave.	St. Dev.	Ave.	St. Dev.
Jan	106.44	76.14	103.69	53.95
Feb	63.66	35.79	51.43	25.46
Mar	63.31	38.62	64.5	39.2
Apr	44.26	25.6	45.58	28.69
May	40.26	38.03	33.82	22.2
Jun	41.15	32.73	27.15	20.89
Jul	36.85	26.36	29.64	32.79
Aug	57.46	81.21	39.26	45.66
Sep	55.22	41.88	37.93	33.55
Oct	100.09	56.94	91.47	60.69
Nov	105.31	48.57	99.06	44.33
Dec	113.44	57.63	110.14	48.22
Average	68.95	46.63	61.14	37.97

rainfall–runoff relationship because in Table 8.7 FL model prediction yields less relative error.

It is obvious that FIS model average relative errors are less than the practically acceptable 10% limit.

TABLE 8.6
Monthly Runoff Data Statistics

| Month | Catchments | | | |
| | Terkos | | Ömerli | |
	Ave.	St. Dev.	Ave.	St. Dev.
Jan	24.66	30.3	48.97	24.05
Feb	49.72	19.71	31.94	17.31
Mar	50.97	17.45	36.23	25.63
Apr	34.93	13.2	17.19	13.31
May	39.15	1.57	9.22	6.65
Jun	36.94	5.95	4.96	4.4
Jul	31.6	7.41	5.02	4.87
Aug	69.33	16.79	6.72	7.83
Sep	48.55	9.43	6.00	8.39
Oct	78.51	30.51	11.37	10.92
Nov	76.93	40.12	18.97	17.26
Dec	85.53	39.46	45.78	22.58
Average	52.24	19.33	20.20	13.60

TABLE 8.7

Terkos Fuzzy Rule Base and Regression Prediction Errors

Observation		Runoff Prediction		Relative Error (%)	
Rainfall (mm)	Runoff (mm)	Fuzzy	Regression	Fuzzy	Regression
5.00	9.20	7.43	6.27	19.24	31.86
9.60	8.30	8.28	6.80	0.24	18.02
12.50	7.50	8.80	7.14	14.77	4.78
14.40	7.80	9.11	7.36	14.38	5.60
15.50	8.00	9.29	7.49	13.89	6.36
20.70	9.60	10.10	8.10	4.95	15.66
24.00	11.00	10.60	8.48	3.64	22.91
28.60	11.40	11.10	9.02	2.63	20.91
29.70	9.80	11.30	9.14	13.27	6.70
32.90	13.20	11.80	9.52	10.61	27.91
35.70	10.70	12.10	9.84	11.57	8.02
36.00	11.30	12.20	9.88	7.38	12.59
39.70	14.90	12.60	10.31	15.44	30.82
41.60	14.30	12.80	10.53	10.49	26.37
43.40	16.60	13.00	10.74	21.69	35.31
48.30	16.10	13.40	11.31	16.77	29.76
51.70	11.80	13.60	11.70	13.24	0.81
54.40	13.90	13.80	12.02	0.72	13.53
60.10	13.70	14.10	12.68	2.84	7.43
69.30	14.20	14.50	13.75	2.07	3.15
90.70	15.30	16.80	16.24	8.93	5.81
96.20	16.60	19.60	16.88	15.31	1.68
96.90	18.00	20.00	16.97	10.00	5.74
103.40	24.80	22.50	17.72	9.27	28.54
107.90	23.20	24.00	18.25	3.33	21.35
109.40	25.80	24.40	18.42	5.43	28.60
127.20	29.10	28.60	20.49	1.72	29.58
137.70	37.50	30.70	21.72	18.13	42.09
149.10	37.70	33.10	23.04	12.20	38.88
154.90	35.40	34.40	23.72	2.82	33.00
164.60	33.50	36.80	24.85	8.97	25.83
178.60	45.70	41.20	26.48	9.85	42.07
327.00	78.80	78.30	43.75	0.63	44.48
			Average	**9.28**	**20.49**

8.6   RAINFALL–GROUNDWATER RECHARGE

Let the system building be the relationship between the rainfall rate i and groundwater recharge g. The first step in building a FIS includes a determination of the system input. The rainfall rate is the input in hydrology for groundwater storage variations to regulate the withdrawal rates, which are the system outputs. Hence, it is significant to model groundwater recharge for the rainfall rate. The rainfall adjectives that can be used in the hydrological sciences are "light," "medium," and "intensive." Output description words for groundwater recharge are determined as *low*, *medium*, and *high*. With these three input and three output fuzzy words as in Figure 8.37, the solution space has $3 \times 3 = 9$ sub-areas. However, hydrologically, only three of them are plausible. The rainfall rate input variable can be related to groundwater recharge output variable through "IF...THEN..." rules.

Rule 1: "IF i is 'light' THEN g is 'low.'"

Rule 2: "IF i is 'medium' THEN g is 'medium.'"

Rule 3: "IF i is 'intensive' THEN g is 'high.'"

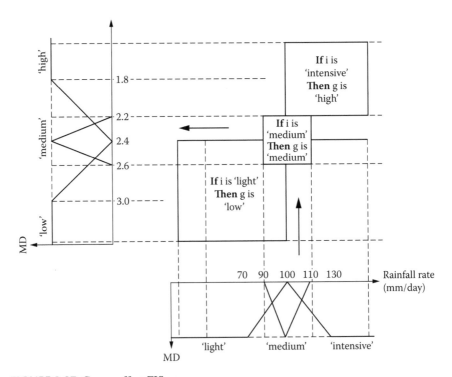

FIGURE 8.37 Cause–effect FIS.

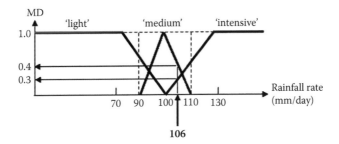

FIGURE 8.38 Rainfall rate above target value.

Figure 8.37 includes fuzzy MFs in the form of triangles and trapeziums, but other shapes, such as bell curves, could also be used (Chapter 3). Narrow (wide) triangles provide tighter (looser) control.

Determination of the groundwater recharge output can be obtained from the rainfall rate. This calculation is time consuming when done manually, but it may take only thousandths of a second when processed by a computer.

Assume something changes in the system that causes the rainfall rate to increase from the target rate of 100 mm/day to 106 mm/day. Hence, the cause chart appears as in Figure 8.38.

The vertical line intersects the "medium" triangle at 0.4 and the "intensive" triangle at 0.3. The next step is to draw output triangles with their height determined by the values obtained above. Because the rainfall rate vertical line of 106 mm/day does not intersect the "light" triangle, it is not drawn. The "medium" change and the "intensive" MFs are drawn because the vertical rainfall rate line intersects them. According to the rainfall rate as 106 mm/day, only R2 and R3 are triggered; hence the consequent MFs are "medium" and "high." Therefore, "medium" ("high") truncation of the MF of g at the 0.4 (0.3) level yields to truncated trapeziums as in Figure 8.39. The result is affected by the widths given to the triangle and trapeziums in Figure 8.38. The "medium" change triangle has a height of 0.4 and the "high" triangle has a height of 0.3 because these were the intersect points for their matching consequent triangles (Figure 8.39). In this figure, O_1 and O_2 are the fuzzy inference sets and they should be defuzzified for a crisp value search. For this purpose, if the centroid method is used, the result becomes 2.62 mm/day.

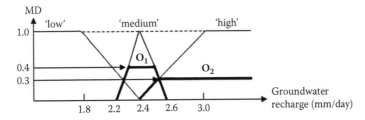

FIGURE 8.39 Control value determination.

TABLE 8.8
Transmissivity Categorization

Transmissivity (m²/day)	Potentiality
$T > 500$	"high"
$500 < T < 50$	"moderate"
$50 < T < 5$	"low"
$5 < T < 0.5$	"weak"
$T < 0.5$	"negligible"

8.7 FUZZY AQUIFER CLASSIFICATION CHART

Basic groundwater aquifer parameters are commonly used for aquifer classification—that is, transmissivity and storativity. The transmissivity is the amount of water that moves through the whole thickness of an aquifer from a unit width under unit hydraulic gradient change. On the other hand, the storage coefficient is the amount of water that can be taken from or stored into the aquifer from the whole thickness from a unit area of the horizontal aquifer area under unit hydraulic gradient change.

The classical hydrogeological assessment of groundwater potentiality is based on transmissivity values given in Table 8.8, where the classifications are crisp and do not have clear meanings at the boundaries. This parameter characterizes the ability of the aquifer to transmit water.

The storage coefficient indicates the ability of an aquifer to store water, which is one of the most important hydraulic properties. The information gathered from the literature (Freeze and Cherry 1989; Todd, 1976) provides the storage coefficient categorization as shown in Table 8.9.

In Figure 8.40 and Figure 8.41, the transmissivity and storativity are presented on a logarithmic scale with 5 and 6 MFs, respectively.

The storage coefficient classifications in Table 8.9 are not adopted as they are in Figure 8.41 but the whole domain of parameter variation is divided into three aquifer types (that is, confined, leaky, and unconfined) in order of increasing storage

TABLE 8.9
Storage Coefficient Categorization

Storativity	Aquifer Type	
$S < 5 \times 10^{-6}$	"fine"	
$5 \times 10^{-6} < S < 5 \times 10^{-5}$	"medium"	
$5 \times 10^{-5} < S < 5 \times 10^{-4}$	"course"	Confined
$5 \times 10^{-4} < S < 5 \times 10^{-3}$	"semi"	
$5 \times 10^{-4} < S < 5 \times 10^{-3}$	"semi" confined	
$5 \times 10^{-3} < S < 5 \times 10^{-2}$	"semi" unconfined	Leaky
$5 \times 10^{-3} < S < 5 \times 10^{-2}$	"semi"	
$5 \times 10^{-2} < S < 5 \times 10^{-1}$	"fine"	Unconfined

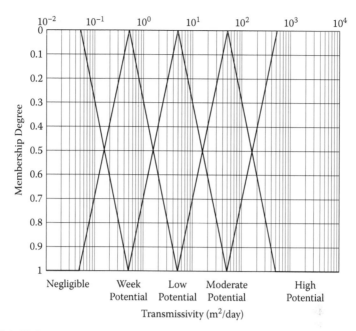

FIGURE 8.40 Transmissivity membership functions.

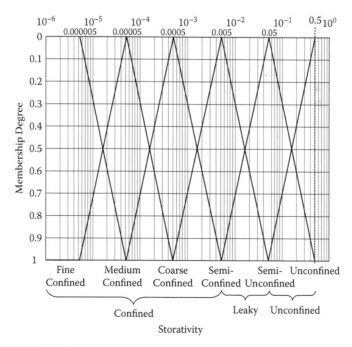

FIGURE 8.41 Storativity membership functions.

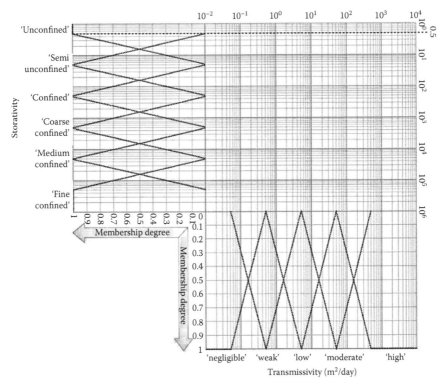

FIGURE 8.42 Fuzzy aquifer classification diagram.

coefficient value. Furthermore, more refined classifications are presented on the basis of aquifer types as "fine confined," "medium confined," "coarse confined," semi-confined," semi-unconfined," and "unconfined," again in order of increasing storativity values. Hence, it is possible to classify any aquifer depending on its storativity value according to three main aquifer types and then sub-types with different MF values.

Although it is possible to classify each aquifer depending on the transmissivity and storativity coefficients separately from the last figures into at least two different categories, Figure 8.42 is more helpful in assessing the joint categorization with the scatter diagram. The scatter domain is now divided into $5 \times 6 = 30$ joint classification sub-domains, each with a degree of potentiality and aquifer storage properties. Accordingly, the fuzzy joint classifications (fuzzy inference) of T and S are given in Table 8.10. The joint specifications in this table are referred to as fuzzy inferences from S and T specifications. The classifications in Table 8.8 are applied as they are in Figure 3.40 through the five MFs. The combination of Tables 8.8 and 8.9 leads to fuzzy rules. Few rules are presented in the following and others have the same format:

R1: "IF T is 'high' and S is 'fine confined' THEN the aquifer has 'high' potential."

R2: "IF T is 'moderate' and S is 'course confined' THEN the aquifer has 'very high' potential."

TABLE 8.10

Hydrogeologic Parameter Fuzzy Inference

		Transmissivity (m²/day)				
		"high"	"moderate"	"low"	"weak"	"negligible"
	"fine"	"H"	"VH"	"H"	"L"	"N"
Confined	"medium"	"VH"	"H"	"L"	"VL"	"N"
	"course"	"VMH"	"VH"	"H"	"M"	"L"
Leaky	"confined"	"VH"	"H"	"M"	"L"	"VL"
	"unconfined"	"H"	"M"	"L"	"VL"	"VML"
Unconfined		"VH"	"L"	"VL"	"VML"	"VMN"

Storativity labels the left vertical axis.

Note: "very much high, "VMH"; "very high," "VH"; "high," "H"; "medium," "M"; "very much low," "VML"; "very low," "VL; "low," "L"; "very much negligible," "VMN"; "very negligible," "VN"; "negligible," "N."

R3: "IF *T* is 'low' and *S* is 'semi confined' THEN the aquifer has 'moderate' potential."

R4: "IF *T* is 'weak' and *S* is 'fine unconfined' THEN the aquifer has 'very very low' potential."

To perform aquifer classification on the basis of the parameters (storativity *S* and transmissivity *T*) herein, the slope matching method developed by Şen (1995) is used with a classical textbook confined aquifer test selected from Todd (1980). A well-penetrating confined aquifer is pumped at a uniform rate of 2500 m³/day. Drawdown during the pumping period is measured in an observation well 60 m away, and the measurements of time (t) and drawdown (s) are listed in Table 8.11.

After convenient Theis-type curve matching with matching point selection, $W(u) = 1.00$, and $u = 1 \times 10^{-2}$, and corresponding $s = 0.18$ m and $r^2/t = 150$ m values, the aquifer parameters are calculated as $T = 1110$ m²/day and $S = 2.06 \times 10^{-4}$, respectively.

The same aquifer test is subjected to slope matching calculation steps and the final results are presented in the last two columns of Table 8.11.

It is obvious that for each time–drawdown measurement after the first reading, there are $u_i W_i(u_i)$, T_i, and S_i values, which change during the whole pumping test. On the other hand, extensive aquifer tests were performed on the Arabian Peninsula in the same aquifer through eight wells (Şen and Wagdani, 2008). Application of the slope method to these tests yielded *S* and *T* values similar to those given in Table 8.11 and they are plotted in Figure 8.43.

Although all the test wells are in the same aquifer, the aquifer parameters differ significantly. Hence, *S* and *T* estimations fall within ranges 1×10^0 to 10^{-7} and 1×10^0 to -2×10^4 m²/day, respectively. According to Tables 8.8 and 8.9, the initial fuzzy results without distinction between the MFs are presented in Table 8.12 based on the number of samples in each joint class.

TABLE 8.11
Slope Matching Calculation Results

Time (day) t, ($^{TM}10^{-3}$)	Drawdown (m)s	λ_i	u_i	$W_i(u_i)$	T_i (m²/day)	S_i ($^{TM}10^{-4}$)
0	0					
0.70	0.20	0.74	0.2517	1.05	889.25	1.70
0.10	0.27	0.37	0.0471	260	1818.31	1.00
1.40	0.30	0.56	0.1633	1.51	941.31	2.40
1.70	0.34	0.46	0.0857	1.98	1109.06	1.80
2.10	0.37	0.36	0.0305	2.72	1386.56	1.00
2.80	0.41	0.42	0.0668	2.24	1037.33	2.10
3.50	0.45	0.35	0.0315	2.74	1171.19	1.40
4.20	0.48	0.34	0.0346	2.80	1104.84	1.80
5.60	0.53	0.33	0.0205	3.00	1086.83	1.40
6.90	0.57	0.28	0.0124	3.51	1193.87	1.10
8.30	0.60	0.32	0.0232	3.09	998.57	2.10
9.70	0.63	0.24	0.0900	3.73	1141.96	11.10
12.50	0.67	0.25	0.0189	3.92	1122.69	3.00
16.70	0.72	0.24	0.0905	3.77	1013.56	17.00
20.80	0.76	0.22	0.0066	4.49	1136.68	1.70
27.80	0.81	0.22	0.0053	4.60	1103.70	1.80
34.70	0.85	0.31	0.0241	3.11	707.97	6.60
41.70	0.90	0.14	0.0001	8.77	1907.42	0.10
55.60	0.93	0.11	0.0004	7.03	1479.05	0.40
69.40	0.96	0.22	0.0063	4.44	900.96	4.40
83.30	1.00	0.18	0.0011	5.68	118.51	1.10
104.20	1.04	0.16	0.0091	6.35	1198.06	12.60
125.00	1.07	0.18	0.0028	5.56	1019.30	4.00
145.80	1.10	0.13	0.0004	7.41	1327.76	0.80
	$T = 1162.70$ m²/day				High potential	
	$S = 3.43 \times 10^{-4}$				Confined aquifer	

Table 8.12 indicates that the aquifer is not homogeneous and the most frequently appearing portion falls under the specification of a "highly potential" leaky unconfined aquifer corresponding to "VMH" joint specification in Table 8.10. The next significant portion of the aquifer the has "moderate potential unconfined" condition, the combination of which gives "low" ("L") combined potential according to Table 8.10.

Table 8.13 presents the complete assessment of the aquifer tests in the same aquifer, including aquifer types, sub-types, and fuzzy MDs for S and T. It is possible to obtain from Figure 8.43 MDs for each data point on the horizontal and vertical axis MFs. Table 8.13 indicates the MDs of the arithmetic averages of each fuzzy inference location from Table 8.12.

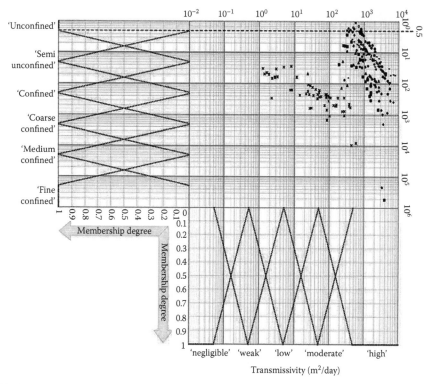

FIGURE 8.43 Field data plot from slope test results.

It is possible to calculate the weighted average of T and S with MDs as weights (Equation 7.2) and the results are $T = 3726.4$ m²/day and $S = 6 \times 10^{-3}$. Hence, according to Tables 8.8 and 8.9, the overall average behavior of the aquifer is "highly potential" and "fine confined" aquifer.

TABLE 8.12
Aquifer Test Results

Aquifer Types			\multicolumn{5}{c}{Transmissivity (m²/day)}				
			"high"	"moderate"	"low"	"weak"	"negligible"
Storativity	Confined	"fine"	2	0	0	0	0
		"medium"	0	0	0	0	0
		"course"	1	3	2	0	0
	Leaky	"confined"	23	12	2	3	0
		"unconfined"	53	20	2	11	0
	Unconfined	"fine"	15	46	0	0	0

TABLE 8.13
Aquifer Classification Details

Aquifer Type	Aquifer Sub-type	Fuzzy Degrees			
		S		T	
		(−)	MD	(m²/day)	MD
	"fine"	1.1×10^{-6}	1.0	7.5×10^3	1.0
Confined	"medium"	2.5×10^{-5}	0.35	1.0×10^4	1.0
	"course"	3.5×10^{-4}	0.15	6.0×10^2	1.0
Leaky	"confined"	3.5×10^{-3}	0.14	1.0×10^2	0.3
	"unconfined"	3.4×10^{-2}	0.14	7.5×10^2	1.0
Unconfined	"fine"	1.8×10^{-1}	0.45	8.0×10^2	1.0

8.8   RIVER TRAFFIC MODEL

This example is adopted from Todorovich and Vukadinovic (1988). The problem of bulk freight transportation includes the loading, transporting, and unloading of gravel along the river. A pusher tug pushes barges; there may be a combination of two, three, or at most four barges. After loading the gravel at the loading point, the tug pushes the barges upstream toward the ports for gravel delivery. When the barges and gravel reach the unloading ports, the dispatcher in charge of traffic control decides on the number of barges that should be left at each port. The barges at a particular port are pushed by the tug for anchoring in the reloading machinery zone. After leaving a certain number of barges in the port, the tug continues to navigate upstream until the entire tow has been left in ports. Then the tug changes direction, heading downward to pick up the empty barges along the river. Again, the dispatcher decides on the number of empty barges that the tug should pick up at each port. After forming a tow of empty barges, the tug continues downstream to the suction-dredging machine for refill. It is assumed that all loading, transportation, and unloading phases take place simultaneously along the river. A dispatcher manages the entire process under uncertainty conditions because the goals, constraints, and consequences of taken actions cannot be perceived with precision. The dispatcher makes subjective but certain decisions in short time durations. The problem becomes more complex, uncertain, and hence fuzzy by the increase in the number of tugs, barges, and ports for unloading. It is not possible to numerically model the entire process because the measurement of different components takes time and in the meantime the process progresses even ahead of the data recording. It is therefore necessary to develop an expert system approach with approximate reasoning by FL algorithms. In this process, the dispatcher endeavors to minimize the tug's waiting time, bearing in mind the fuzzy variable "waiting time to unload." Another fuzzy variable in the process is "number of barges left." With these two fuzzy variables, the dispatcher makes certain fuzzy rules to make the final decision on the number of barges left in a specific port. For example, one of the rules might be:

"IF waiting time to unload is 'large' THEN the number of barges left is 'small.'"

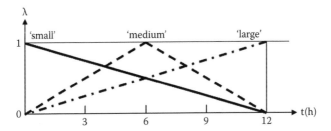

FIGURE 8.44 Waiting time to upload MFs.

It is assumed that the dispatcher divided the variation domains of two fuzzy variables into three fuzzy words—namely, *small*, *average*, and *large*.

Endeavoring to form an empty tow for the tug navigating downstream, the dispatcher considers a third fuzzy variable as "number of empty barges in the port," again with the same fuzzy sub-wordings. Another fuzzy variable with the same division is "number of empty barges taken by tug." As in the case of the upstream tug, the dispatcher applies certain rules to arrive at a final decision regarding the number of empty barges to be taken by the tug. Figure 8.44 indicates the MFs for fuzzy variable "waiting time to unload." The number of barges left is adopted as a positive integer number y between 0 and 4 inclusive. Accordingly, the fuzzy singletons in "small," "medium," and "large" number of barges are labeled respectively by fuzzy sets A, B, and C, which are given as follows:

$$A = \left\{ \frac{1}{0} + \frac{0.75}{1} + \frac{0.5}{2} + \frac{0.25}{3} + \frac{0}{4} \right\}$$

$$B = \left\{ \frac{1}{0} + \frac{0.50}{1} + \frac{1.0}{2} + \frac{0.50}{3} + \frac{0}{4} \right\}$$

and

$$C = \left\{ \frac{0}{0} + \frac{0.25}{1} + \frac{0.5}{2} + \frac{0.75}{3} + \frac{1}{4} \right\}$$

Let us assume that the navigation takes place between a downstream station D, middle-stream port M, and an upstream port U. The tug can take empty barges from port D and then that same tug completes its tow in port M and further navigates toward the dredging machine. The dispatcher endeavors to make decisions that will enable the tug going downstream to form a complete barge tow as quickly as possible. The following questions come to the dispatcher's mind:.

1. How many empty barges should the tug take from the downstream port? The fuzzy variable is "number of empty barges in D."
2. What is the number of empty barges waiting at port M? The fuzzy variable is "number of empty barges in M."

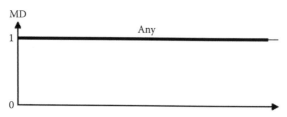

FIGURE 8.45 MF of "any."

3. Both variables can assume "small," "medium," and " large" integer numbers according to the following sequence of labels A_1, B_1, and C_1, respectively:

$$A_1 = \left\{ \frac{1}{0} + \frac{1}{1} + \frac{0.67}{2} + \frac{0.33}{3} + \frac{0}{4} \right\}$$

$$B_1 = \left\{ \frac{0}{0} + \frac{0.5}{1} + \frac{1}{2} + \frac{0.5}{3} + \frac{0}{4} \right\}$$

and

$$C_1 = \left\{ \frac{0}{0} + \frac{0.33}{1} + \frac{0.67}{2} + \frac{1}{3} + \frac{1}{4} \right\}$$

In the navigation process, *any* number of barges or "any" amount of waiting time to unload may also play a role. It is therefore necessary to define an MF for the fuzzy word *any*. The MF of every element belonging to fuzzy set " any" equals 1; Figure 8.45 shows the MF of "any."

Vukadinovic and Todorovich (1994) developed two approximate reasoning algorithms. The first one calculates the number of barges that the tug should leave in port M, and the second one calculates the number of empty barges it should take in D. The algorithm that calculates the number of barges to be left in M consists of the following rules:

R1: "IF waiting time to unload in M is 'large' AND waiting time to unload in D is 'large' OR 'average' THEN the number of barges left in M is 'average'"

OR

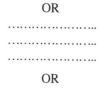

OR

R5: "IF waiting time to unload in M is 'small, AND waiting time to unload in D is average' OR 'small' THEN the number of barges left in M is 'average.'"

Figure 8.46 represents graphically the approximate reasoning by max-min composition to calculate the number of barges remaining in M.

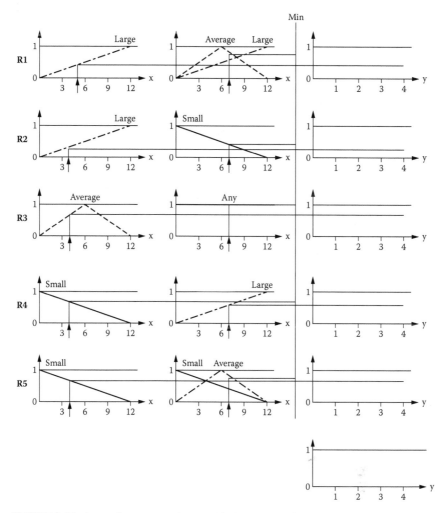

FIGURE 8.46 Approximate reasoning graphical representation.

The following rules are valid for the approximate reasoning to calculate the number of empty barges that are picked in D:

R1: "IF the number of empty barges in D is 'large' AND the number of empty barges in M is 'any' THEN the number of picked up in D is 'large.'"

TABLE 8.14
Number of Barges Left Based on Approximate Reasoning Model

	0	1	2	3	4	5	6	7	8	9	10	1	12
0	2	2	2	2	2	2	2	2	1	1	1	1	1
1	2	2	2	2	2	2	2	2	1	1	1	1	1
2	2	2	2	2	2	2	2	2	2	1	1	1	1
3	2	2	2	2	2	2	2	2	2	2	1	1	1
4	2	2	2	2	2	2	2	2	2	2	2	1	1
5	2	2	2	2	2	2	2	2	2	2	2	2	2
6	2	2	2	2	2	2	2	2	2	2	2	2	2
7	2	2	2	2	2	2	2	2	2	2	2	2	2
8	3	3	2	2	2	2	2	2	2	2	2	2	2
9	3	3	3	2	2	2	2	2	2	2	2	2	2
10	3	3	3	3	2	2	2	2	2	2	2	2	2
11	3	3	3	3	3	2	2	2	2	2	2	2	2
12	3	3	3	3	3	2	2	2	2	2	2	2	2

R4: "IF the number of empty barges in D is 'small' AND the number of empty barges in M is 'small' THEN the number of picked up in D is 'average.'"

The rule R4 means that the tug should wait in D until the number of empty barges becomes an "average" number of barges. Vukadinovic and Todorovic (1994) tested these approximate reasoning algorithms on a large number of numerical examples for Belgrade as D and Pancevo as M ports. Tables 8.14 and 8.15 represent, respectively, the number of barges left based on the approximate reasoning algorithm and

TABLE 8.15
Number of Barges Left Determined by Dispatcher

	0	1	2	3	4	5	6	7	8	9	10	1	12
0	2	2	2	2	2	2	2	1	1	1	1	1	1
1	2	2	2	2	2	2	2	1	1	1	1	1	1
2	2	2	2	2	2	2	2	2	1	1	1	1	1
3	2	2	2	2	2	2	2	2	2	1	1	1	1
4	2	2	2	2	2	2	2	2	2	2	1	1	1
5	2	2	2	2	2	2	2	2	2	2	2	2	1
6	3	2	2	2	2	2	2	2	2	2	2	2	2
7	3	3	2	2	2	2	2	2	2	2	2	2	2
8	3	3	3	2	2	2	2	2	2	2	2	2	2
9	3	3	3	3	2	2	2	2	2	2	2	2	2
10	3	3	3	3	3	2	2	2	2	2	2	2	2
11	3	3	3	3	3	3	2	2	2	2	2	2	2
12	3	3	3	3	3	3	3	2	2	2	2	2	2

the number left based on the dispatcher survey for different waiting times in Belgrade and Pancevo.

REFERENCES

Freeze, R.A. and Cherry, J.A. 1989. *Groundwater.* Prentice-Hall International, Inc., London, 605 pp.

Horton, R.E. 1940. An approach towards a physical interpretation of infiltration capacity. *Trans. Am. Geophys. Un.,* 20(IV): 693–711.

Kadioglu, M. and Şen, Z. 2001. Monthly precipitation–runoff polygons and mean runoff coefficients. *Hydrolog. Sci. J.* 46(1): 3–11.

Keskin, M.E., Terzi, Ö., and Taylan, D. 2004. Fuzzy Logic model approaches to daily pan evaporation estimation in western Turkey. *Hydrological Sciences Journal,* 49(6): 1001–1010.

Kundzewicz, Z.W. and Napieorkowski, J.J. 1986. Nonlinear models of dynamic hydrology. *Hydrolog. Sci. J.* 37: 163–185.

Ross, J.T. 1995. *Fuzzy Logic with Engineering Applications.* McGraw-Hill Inc, New York, 593 pp.

Sugeno, M. 1977. Fuzzy measures and fuzzy integrals: a survey, In Gupta, M., Saridis, G.N., and Gaines, B.R., Eds., *Fuzzy Automata and Decision Processes,* North-Holland, New York, pp. 329–346.

Şen, Z. 1995. *Applied Hydrogeology for Engineers and Scientists.* CRC Press/Lewis Publishers, Boca Raton, FL, 465 pp.

Şen, Z. 2001. Bulanık Mantık ve Modelleme İlkeleri (Fuzzy Logic and Modelling Principles), Bilgi, Sanat ve Kültür Basimevi, Istanbul (in Turkish).

Şen, Z. and Wagdani, E. 2008. Aquifer heterogeneity determination through the slope method. *Hydrolog. Processes* 22(12): 1788–1795.

Todd, D.K. 1976. *Ground Water Hydrology. (2nd ed.).* John Wiley & Sons, New York.

Todd, D.K. 1980. Groundwater Hydrology. John Wiley & Sons. New York. 535 pp.

Todorovich, D., and Vukadivich, K., 1988. Taffic Control and Transport Planning. Kluwer Academic Publishers, Boston, 383 pp.

Vukadinovic, K. and Todorovich, D. 1994. A Neuro and Fuzzy Control of Inland Water Transportation with Applications to Other Models of Transportation. Ph.D. thesis, Faculty of Transport and Traffic Engineering, University of Belgrade (in Serbian).

9 Fuzzy Water Resources Operation

9.1 GENERAL

Hydrologic processes require the application of a water budget, which is explicitly written in the form of an equation between inputs and outputs but it is an approximate "balance" of water. Similar "approximations" are also valid in all water resources management, operation, and processing, which makes the FL principles, rules, and inference system applicable. The very word *balance* in water budget studies implies that the input is "almost" equal to output, which indicates a 45° fuzzy proportionality line.

9.2 FUZZY WATER BUDGET

One of the preliminary significant principles in any science is the mass conservation equation, which is also known in the hydrology domain as the water balance, continuity or water budget equation. In a hydrologic system, it expresses the relationship among three linguistic variables—namely, the total input I, the total output O, and the change in the storage $\pm \Delta S$, as

$$I - O = \pm \Delta S \tag{9.1}$$

Its practical application requires specifications of time and area domains. In conventional applications, the time interval is finite as in hour, day, month, season, or year, and the area is mostly the drainage basin area. The term on the right-hand side refers to time variation in the storage.

After appropriate decisions on the time interval and area, Equation (9.1) aids in calculating one of the three variables, provided that other two variables are known. Most often, its practical application is considered a rather mechanical procedure without considering the logical background of the principle itself. To obtain more insight into the water budget equation, it is necessary to interpret this principle linguistically and graphically as in Figure 9.1 where the total input and output domain is divided crisply into three distinctive regions. Because the input and output variables have absolute physical quantities, their common domain of variability is confined only to the first quadrant of the Cartesian coordinate system.

1. In the FL sense, if the total input is "close" to the total output, then the water "balance" appears as an almost 45° line that passes through the origin.

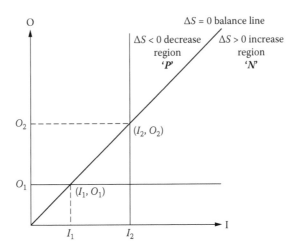

FIGURE 9.1 Crisp water balance principle representation.

However, in the case of absolute equality, this is a crisp line. Even infinitesimally small deviations from this line pose "imbalances."

2. Below the 45° balance line, the storage change is positive ($\Delta S > 0$), and therefore it can be labeled as an increment in the storage, which implies mathematically that $I > O$; however, CL does not tell "how much" the "increase" is in words such as *small*, *medium*, *big*, etc.

3. Above the 45° balance line, the storage change is negative ($\Delta S < 0$), and therefore it can be labeled as a decrement in the storage, which implies mathematically that $I < O$ without any fuzzy specification about the "decrease".

The representation of these regions by CL can be achieved by attaching the increment (positive) "P" region and decrement (negative) "N" region.

For further explanation, let us concentrate on constant I or O levels with locations of (I_1, O_1) and (I_2, O_2) in the joint variation domain of I and O as shown in Figure 9.1 and on imaginary cross sections through each direction. The resulting profiles appear as in Figure 9.2, where each line has one point from the balance line. Herein, the characteristic value (CV) of "N" is taken as 1. This is an ideal situation in terms of the nature of the hydrological events, whereas in engineering applications, one cannot be so decisive, and there are always approximations. For instance, in reservoir operation and management, rather than such crisp rules, human thinking soft rules find continuous application because they subsume the linguistic information of expert views.

It is possible to expand the water budget principle into an FIS by pondering the possible relationships between the three variables on the basis of fuzzy subsets for each variable. Hence, the first step is to fuzzify the variables (I and O) into vague boundary sets as shown in Figure 9.3, which is similar to Figure 9.2 but with fuzzy demarcations. Herein, "P" and "N" have different but overlapping MFs.

If the input and output linguistic variables have each three MFs as "small" ('SI' and 'SO'), "medium" ('MI' and 'MO'), and "high" ('HI' and 'HO'), respectively, then

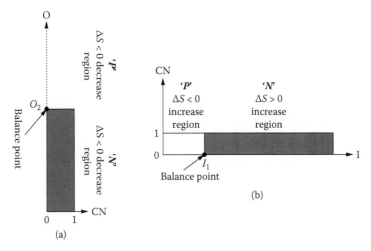

FIGURE 9.2 Cross sections and crisp sets: (a) I_2 constant level and (b) O_1 constant level.

each one of the joint sub-area implies the "IF ... THEN ..." rule and the antecedent part with "ANDing" conjunctive. For simplification, the MFs are considered triangles and trapeziums. However, other type of MFs can also be adopted, as discussed in Chapter 3. Hence, nine mechanical water balance (antecedent) predicate rules are shown in Figure 9.4 and explicitly are:

R1: "IF input is 'SI' AND output is 'SO' THEN ..."

R2: "IF input is 'SI' AND output is 'MO' THEN ..."

R3: "IF input is 'ST' AND output is 'HO' THEN ..."

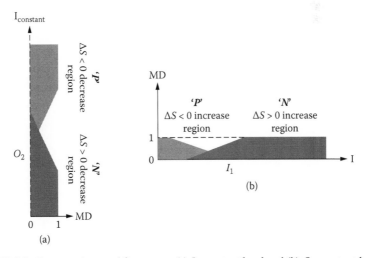

FIGURE 9.3 Cross sections and fuzzy sets: (a) I_2 constant level and (b) O_1 constant level.

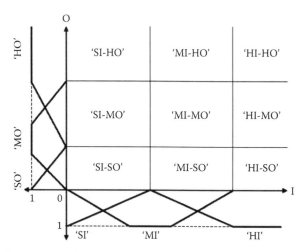

FIGURE 9.4 Fuzzification of input-output domain.

R4: "IF input is 'MI' AND output is 'SO' THEN . . ."

R5: "IF input is 'MI' AND output is 'MO' THEN . . ."

R6: "IF input is 'MI' AND output is 'HO' THEN . . ."

R7: "IF input is 'HI' AND output is 'SO' THEN . . ."

R8: "IF input is 'HI' AND output is 'MO' THEN . . ."

R9: "IF input is 'HI' AND output is 'HO' THEN . . ."

In Figure 9.4 the common sub-areas are shown in terms of input fuzzy wordings. The consequent part of each rule is left empty because the consequence effects are not discussed yet.

For water budget fuzzification, the final step is to locate the "storage change" linguistic variable MFs as shown in Figure 9.5. Herein, "BNS," "NS," "ZS," "PS," and "BPS" indicate "big negative small," "negative small," "zero small," "positive small," and "big positive small" fuzzy sets, respectively.

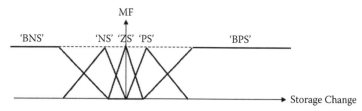

FIGURE 9.5 Storage change fuzzy subsets.

TABLE 9.1

FAM of Water Balance

		Input		
		"ST'	"MI"	"HI"
Output	"SO"	"ZS"	"PS"	"BPS"
	"MO"	"NS"	"ZS"	"PS"
	"HO"	"BNS"	"NS"	"ZS"

It is now time to attach to each one of these nine sub-domains one of the suitable storage change fuzzy sets. At this stage, the expert view about the water budget principle is most needed, and the consequent parts in the rule bases depend significantly on the hydrological regime of the area concerned. It is different in arid regions than in humid areas.

The configuration of the three-variable fuzzy subsets in this Table 9.1 in filling the consequent part of the water budget principles as:

R1: "IF input is 'SI' AND output is 'SO' THEN storage change is 'ZS.'"

R2: "IF input is 'SI' AND output is 'MO' THEN storage change is 'NS.'"

R3: "IF input is 'SI' AND output is 'HO' THEN storage change is 'BNS.'"

R4: "IF input is 'MI' AND output is 'SO' THEN storage change is 'PS.'"

R5: "IF input is 'MI' AND output is 'MO' THEN storage change is 'ZS.'"

R6: "IF input is 'MI' AND output is 'HO' THEN storage change is 'NS.'"

R7: "IF input is 'HI' AND output is 'SO' THEN storage change is zero 'BPS.'"

R8: "IF input is 'HI' AND output is 'MO' THEN storage change is 'PS.'"

R9: "IF input is 'HI' AND output is 'HO' THEN storage change is 'ZS.'"

It is possible to write down the water budget rule base in the form of an FAM as in Table 9.1.

9.3 DRINKING WATER CONSUMPTION PREDICTION

Drinking water sources contain various natural and anthropogenic contaminants. A "low" concentration of these contaminants is not harmful to human health and removal of all such contaminants is costly. Provided that the contaminant levels are less than specific standards, they do not present harmful effects on human health. In general, physical human activities such as physical activity, body weight, and body temperature play a significant role in drinking water consumption rates. Most often, human activity variables are given in crisp numerical interval classifications for water consumption calculations.

Although there are many statistical or stochastic methods for modeling quality concentrations or drinking water consumption rates, they are based on crisp interval values attached to linguistic specifications. Given the standard crisp values, the water is regarded as "uncontaminated" as long as they fall within crisply defined intervals. In practice, the entire range of plausible values can be subdivided into linguistic subsets by fuzzy words such as *low*, *medium*, *high*, *good*, *moderate*, *poor*, *brackish*, etc. In everyday practice, these sub-ranges are viewed as not overlapping with each other. However, in any fuzzy modeling, they are considered overlapping such that with the deterioration of one range, the improvement of the adjacent one starts to take over (Kiska et al., 1985; Kosko, 1987; Mamdani, 1974; Ross, 1995; Şen, 2001; Zadeh, 1965).

Both variability and uncertainty in drinking water consumption encompass a multiplicity of concepts and the precise meaning of these terms varies across disciplines (Kundu et al., 2004). These concepts depend on issues that distinguish between inherent physical or natural characteristics and limitations of knowledge or understanding, thus leaving the planner with uncertain, incomplete, and vague information—that is, fuzzy data. The uncertainty aspects in drinking water have already been explained by the EPA (1997). Uncertainties are natural and refer to observed or measurable differences attributable to diversity in a population; for instance, members of a population exhibit variability with respect to their weight. Imprecision is defined as a degree of uncertainty across an exposed population due to inter-subject differences in exposure conditions, rates of intake depending on environmental and body temperatures, inhalation rates (physical activity) per unit body mass, uptake fraction, retention characteristics, bio-transformation, and sensitivity (Raucher et al., 2000).

While future water demand depends as much on consumer preference (or individual unit consumption) as it does on population, much less attention has been given to consumer preferences, which can be determined by market purchase analysis and vary from place to place with cultural, environmental, and other features. In cases of crisp data availability, regression techniques can be used to relate consumer preference (such as drinking water consumption) to specific independent variables. Future preference trends will result from the intersection between the introduction of new goods and services and the changes in the means to purchase and time to enjoy them. Any regression analysis requires a set of assumptions such as the linearity, normality, and independence of errors; homoscedasticity; etc., which are not satisfied most frequently in practice (Benjamin and Cornell, 1973). In addition, regression techniques are not capable of digesting linguistic fuzzy data. In particular, drinking water consumption variables are mostly linguistic, and therefore regression approaches cannot be employed easily in their treatment. This opens up an avenue for the application of fuzzy modeling rather than probabilistic, statistical, or stochastic techniques because the latter require numerical data only. This is the reason why the fuzzy approach is suggested, developed, and applied to drinking water quality variations in this section.

Logically, drinking water consumption depends not only on the measured qualities, but also vaguely on social variables such as human weight, activity, and temperature. In practice, all these variables are presented in tabular form with crisp
interval classification. On the other hand, all studies on drinking water consumption
data are available in the form of short-term surveys with uncertainties. Most often,
it is not possible to have enough numerical data but rather supplementary linguistic
data such as the consumer"s opinion. There is no other approach than the FL and system modeling to treat such linguistic (verbal) data sources. This may be a source of
uncertainty in the consumption rate estimations because of the subjective nature of
survey techniques. However, they include general tendencies, at least for the drinking water consumption rates.

9.3.1 DATA AND RULE BASE SETS

The water consumption of an individual is directly proportional to body weight.
Crisp logically: as body weight increases, the water consumption per unit weight
decreases. Hence, there is an inverse relationship between these two variables. For
practical classifications, each variability range is divided into a corresponding number of non-overlapping but adjacent intervals and most often each interval is specified by a representative linguistic word. The relationship between body weight and
per-kilogram drinking water consumption at 370°C is given in Table 9.2.

The first (second) row in Table 9.2 implies that for the first (second) 10 kg of
body weight, 100 mL/kg (50 mL/kg) drinking water consumption is necessary. If
the body weight is more than 20 kg, then consumption is 20 mL/kg. According
to the classifications given in this table, an individual who weighs 60 kg should
require $(100 \times 10) + (50 \times 10) + (20 \times 40) = 2300$ mL/day. Table 9.2 implies that
babies need more water than small children and adults. Furthermore, under normal
conditions at 1 atmosphere pressure and 37°C of body temperature, 1400 mL/day
of the 2300 mL/day water is released through urine, 100 mL/day through transpiration, and another 100 mL/day by other means. The remaining 700 mL/day is lost by
evaporation through the breathing system. However, due to high temperatures in hot
weather, water losses increase. In extremely hot weather, the water loss may reach
1.5 to 2.0 L/hour through transpiration and there is not doubt that this causes high
rates of water loss and the consequent dehydration effect.

On the other hand, physical training gives rise to water loss by two means. First,
due to high breathing rates, the human body"s ventilation rate increases and the rise

TABLE 9.2
Body Weight and Water Consumption

Body Weight (kg)	Water Consumption (mL/kg)
0–10	100
10–20	50
>20	20

TABLE 9.3

Water Losses

	Normal Temperature (mL/day)	High Temperature (mL/day)	Extended Heavy Training Periods (mL/day)
Hidden loss (skin)	350	350	350
Respiration	350	250	650
Urine	1400	1200	500
Sweat	100	1400	5000
Excrement	100	100	100
Total	**2300**	**3300**	**6600**

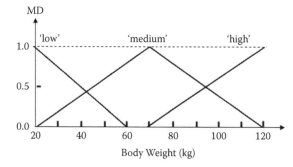

FIGURE 9.6 MFs of body weight (kg).

in body temperature causes transpiration increments; hence, approximately 2.5 mL water is lost daily. Table 9.3, shows the water losses of an individual who weighs 60 kg under normal temperatures; It also shows the relationship between the air temperature and extended heavy physical training periods.

The first factor that affects daily drinking water consumption is body weight. Based on census and demographic details, this variable is divided, herein into three fuzzy subsets ("low," "medium," and "high"), each of which is shown as a triangular MF in Figure 9.6, which represents body weight range between 0 and 120 kg.

The physical activity as the second affecting factor on drinking water consumption is represented by four fuzzy subsets. The data presented in Table 9.4 by Thompson et al. (2004) is used for this purpose. The physical activity levels represent 24-hour

TABLE 9.4

Physical Activity Levels

Chair-bound or bad-bound	1.0–1.6
Seated work with no option of moving around and little or no strenuous leisure activity	1.4–2.0
Standing work (e.g., housewife, shop assistant)	1.8–2.5
Strenuous work or highly active leisure	2.0–2.5

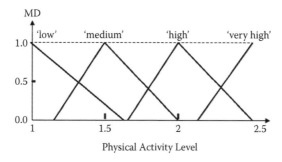

FIGURE 9.7 MFs of physical activity level (multiples of BMR).

averages. Figure 9.7 indicates the approximate situations of four fuzzy subsets as "low," "medium," "high," and "very high". Note that the range between 0 and 2.5 is given as multiples of basic metabolic rate (BMR).

The third factor that affects daily drinking water consumption is the air temperature. Herein, the air temperature is considered to vary between 0°C and 40°C. The fuzzy subsets of the temperature as "low," "medium," and "high" are given in Figure 9.8.

According to the *Exposure Factors Handbook* (USEPA, 1996), water consumption volumes for adults are adopted as varying between 416 and 3780 mL/day. In such a model, per-person daily water consumption is considered to vary from 500 to 4000 mL/day. As shown in Figure 9.9, eight fuzzy subsets are considered for the water consumption amount. These are "low-low" ("LL"), "low" ("L"), "highly low" ("HL"), "low-medium" ("LM"), "medium-medium" ("MM"), "medium" ("M"), "high" ("H"), and "very-high" ("VH"). Figure 9.9 constitutes the consequent part of the fuzzy rule bases.

To develop the Mamdani FIS model for drinking water consumption prediction, the body weight, physical activity, and temperature constitute the antecedent variables with three, three, and four fuzzy subsets, respectively. This implies, in general, that there are $3 \times 3 \times 4 = 36$ mechanical rules, each attached to one of the convenient eight fuzzy subsets for the drinking water consumption variable. Hence, some of these 36 rules may have the same consequent fuzzy subsets. The consequent part

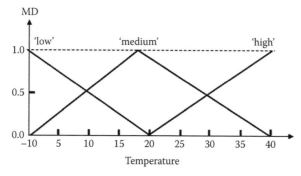

FIGURE 9.8 MFs of temperature (°C).

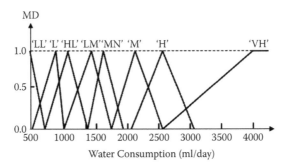

FIGURE 9.9 MFs of water consumption amount (mL/day).

fuzzy subsets of drinking water consumption are allocated according to the expert view of the authors and some other specialists in the study topic. Their consensus views are taken as the final decision in establishing fuzzy rule consequent parts in light of 36 different alternatives in the antecedent part with three variables. Hence, prior to actual data usage, the FIS model is obtained as a collection of "IF . . . THEN . . ." rules in Table 9.5. Such a fuzzy system is very flexible and can digest the imprecise type of information. In this table, the second, third, and fourth columns include the combinations of input variable fuzzy subsets (antecedent part), and the fifth column exposes the corresponding fuzzy rule inferences as the consequent part.

The application of actual data to such a fuzzy system with 36 rules might not trigger some of these rules. Hence, untriggered rules are not relevant for the problem at hand and should be dismissed from further consideration. If there are significant and regression-type relationships between the antecedent variables and the consequent drinking water consumption amounts, then many of the rules may not be triggered. Otherwise, for very scattered data situations, almost all the rules may be triggered at different frequencies. To appreciate this point, the body weight, activity level, air temperature, and drinking water consumption data presented by Istanbul University, Çapa Medicine Faculty is given in Table 9.6. Although it is possible to fit a multi-variable regression model in terms of body weight, activity level, and temperature as independent, and water consumption as the dependent variable such as avenue has not be considered due to the scatter diagrams in Figure 9.10a. Figure 9.10b shows the scatter diagrams of water consumption versus activity level and temperature. Due to such high dispersion, it is not possible to employ a regression approach with restrictive assumptions as explained in previous sections.

However, the FIS approach is quite suitable for dealing with such scatter diagrams. The more the dispersion in the scatter diagram, the more the number of rules triggered. The data provided by Rajkumar et al. (1999) are employed for the first application of the fuzzy model suggested in this section. For this purpose, the fuzzy rules in Table 9.5 are used with the antecedence variables in Tables 9.3 and 9.4; and subsequently in Table 9.7, water consumption predictions are presented for each rule. Figures 9.5, 9.6, and 9.7 show that an entrance with body weight of 70 kg, physical level of 1.76 BMR, and air temperature of 20°C does not trigger rule numbers 1, 4, 9, 12, 25, 28, 33, and 36. The triggered rules lead to the consequent (water consumption) part according to Mamdani (1974) derivation (Chapter 7), which takes into consideration the "minimum"

TABLE 9.5
Drinking Water Consumption Rule Base

Rule No.	Antecedent Parts			Consequent Parts
	Body Weight (kg)	Activity (bbr)	Temperature (°C)	Water Consumption (mL/day)
1	"low"	"low"	"low"	"low-low"
2	"low"	"medium"	"low"	"low-low"
3	"low"	"high"	"low"	"low"
4	"low"	"very-high"	"low"	"high-low"
5	"low"	"low"	"medium"	"high-low"
6	"low"	"medium"	"medium"	"high-low"
7	"low"	"high"	"medium"	"low-medium"
8	"low"	"very-high"	"medium"	"low-medium"
9	"low"	"low"	"high"	"medium"
10	"low"	"medium"	"high"	"high"
11	"low"	"high"	"high"	"high"
12	"low"	"very-high"	"high"	"very-high"
13	"medium"	"low"	"low"	"low"
14	"medium"	"medium"	"low"	"low"
15	"medium"	"high"	"low"	"high-low"
16	"medium"	"very-high"	"low"	"low-medium"
17	"medium"	"low"	"medium"	"low-medium"
18	"medium"	"medium"	"medium"	"low-medium"
19	"medium"	"high"	"medium"	"medium-medium"
20	"medium"	"very-high"	"medium"	"medium-medium"
21	"medium"	"low"	"high"	"medium"
22	"medium"	"medium"	"high"	"high"
23	"medium"	"high"	"high"	"very high"
24	"medium"	"very-high"	"high"	"very-high"
25	"high"	"low"	"low"	"low"
26	"high"	"medium"	"low"	"low
27	"high"	"high"	"low"	"high-low"
28	"high"	"very-high"	"low"	"very-high"
29	"high"	"low"	"medium"	"low-medium"
30	"high"	"medium"	"medium"	"low-medium"
31	"high"	"high"	"medium"	"low-medium"
32	"high"	"very-high"	"medium"	"low-medium"
33	"high"	"low"	"high"	"medium-medium"
34	"high"	"medium"	"high"	"high"
35	"high"	"high"	"high"	"high"
36	"high"	"very-high"	"high"	"high"

TABLE 9.6

Data by Istanbul University, Çapa Faculty of Medicine

No.	Body Weight (kg)	Activity Level (bbr)	Temperature (°C)	Consumption (mL/day)	Prediction (mL/day)	Relative Error (%)
1	36.3	1.49	19.3	1184	1180	0.34
2	43.8	1.07	20.2	1332	1350	1.33
3	43.4	1.11	21.7	1480	1480	0.00
4	42.9	1.85	20	1628	1610	1.11
5	50.9	1.8	20.2	1776	1700	4.28
6	61	1.96	20.2	1924	1910	0.73
7	68	1.9	21.2	2072	2070	0.10
8	69.1	2	21.7	2220	2230	0.45
9	70.1	1.67	37.6	2368	2340	1.18
10	75.2	1.74	37.6	2516	2580	2.48
11	84.3	2.17	28.4	2664	2760	3.48
12	88.3	2.43	36.6	2812	2950	4.68
13	88.9	1.78	40	2960	3060	3.27
14	94.3	2.01	40	3108	3430	9.39
15	96.9	2.5	38.6	3256	3260	0.12
16	101.4	2.43	40	3404	3390	0.41
					Average	**2.08**

of the antecedent MDs in triggering rules. Because in many triggered rules the minimum of the three antecedent parts is equal to zero, there will not be any consequent value for such rules. That is, even in the case of a triggered rule, if its minimum MD on the antecedent part is equal to zero, there will not be any drinking water consumption prediction for the corresponding consequent part. The defuzzification procedure is achieved according to the centroid principle, as explained in Chapter 7.

The complete set of numerical results as a drinking water consumption prediction is presented in Table 9.7 along with relative error percentages for each rule. The maximum relative error reaches 11 percent in two cases but the average error is 3.4 percent, which is practically below the well-accepted limit of 5 percent. Hence, the proposed FIS model for the water consumption prediction is reliable, and such an error level can be easily accepted in any engineering control work. In a previous study, Rajkumar et al. (1999) obtained an average error level of 15 percent; and thus the model developed herein is more refined and acceptable.

To check the reliability of the model developed herein using Rajkumar et al.'s (1999) data, an independent data set was obtained from the Çapa Medicine Faculty, Istanbul University, Turkey, as presented in Table 9.6. The application of the same rule set in Table 9.7 to this data yielded water consumption predictions as shown in Table 9.6. It is obvious that the measured and predicted water consumptions are very close to each other, with less than 1 percent relative error. This shows the validity of the fuzzy rule set in Table 9.7 for water consumption predictions provided that there is body weight, physical activity, and temperature data.

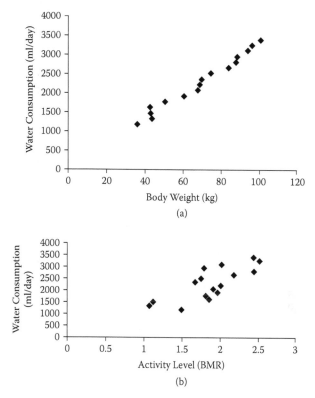

FIGURE 9.10 Scatter diagrams of Istanbul data.

Finally, Figure 9.11 shows observed versus predicted water consumption amounts. Both Rajkumar et al. (1999) and Istanbul University data fall around the 45° straight line with acceptably small deviations. Because the overall deviations from this straight line are less than 5 percent for both data cases, the fuzzy model is acceptable for practical drinking water consumption rate predictions.

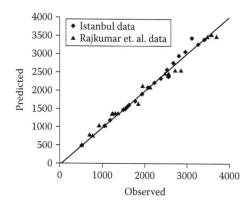

FIGURE 9.11 Observation versus prediction.

TABLE 9.7

Fuzzy Model Results

Rule No.	Body Weight	Activity	Temperature	Data	Model	R.E. (%)
1	"low"	"low"	"low"	502	502	0
2	"low"	"medium"	"low"	515	511	1
3	"low"	"high"	"low"	711	771	8
4	"low"	"very-high"	"low"	1036	1030	1
5	"low"	"low"	"medium"	1070	1040	3
6	"low"	"medium"	"medium"	1074	1040	3
7	"low"	"high"	"medium"	1232	1330	7
8	"low"	"very-high"	"medium"	1355	1360	0
9	"low"	"low"	"high"	1942	2130	9
10	"low"	"medium"	"high"	2559	2470	3
11	"low"	"high"	"high"	2711	2560	6
12	"low"	"very-high"	"high"	3565	3530	1
13	"medium"	"low"	"low"	762	769	1
14	"medium"	"medium"	"low"	766	759	1
15	"medium"	"high"	"low"	924	1040	11
16	"medium"	"very-high"	"low"	1240	1370	9
17	"medium"	"low"	"medium"	1380	1380	0
18	"medium"	"medium"	"medium"	1384	1380	0
19	"medium"	"high"	"medium"	1542	1540	0
20	"medium"	"very-high"	"medium"	1572	1580	1
21	"medium"	"low"	"high"	2123	2100	1
22	"medium"	"medium"	"high"	2466	2460	0
23	"medium"	"high"	"high"	3689	3490	5
24	"medium"	"very-high"	"high"	3479	3490	0
25	"high"	"low"	"low"	762	771	1
26	"high"	"medium"	"low"	766	762	1
27	"high"	"high"	"low"	924	1040	11
28	"high"	"very-high"	"low"	1313	1380	5
29	"high"	"low"	"medium"	1380	1380	0
30	"high"	"medium"	"medium"	1380	1380	0
31	"high"	"high"	"medium"	1380	1380	0
32	"high"	"very-high"	"medium"	1379	1380	0
33	"high"	"low"	"high"	1833	1640	11
34	"high"	"medium"	"high"	2559	2410	6
35	"high"	"high"	"high"	2711	2560	6
36	"high"	"very-high"	"high"	2856	2560	10
					Averages	3.4

9.4   FUZZY VOLUME CHANGE IN RESERVOIR STORAGE

The volume elevation curve of any dam appears in the form of Figure 9.12, where after a certain volume value, the elevation increase ceases. The filling of a dam is a nonlinear procedure such that its filling must be smooth with no damage. The "smaller" ("bigger") the elevation h, the smaller (bigger) the volume v. In the dry stage of a dam, the reduction rate of elevation should be reduced nonlinearly such that its minimum level will be smooth without any damage. The "smaller" the elevation h, the "bigger" the elevation reduction rate. as shown in Figure 9.12.

The two state variables in a dam filling are the elevation and the volume. The control output is the runoff r that will be exerted on the dam, depending on the situation. The following relationships represent a dam filling system:

$$v_{i+1} = v_i + r_i \qquad (9.2)$$

and

$$h_{i+1} = h_i + v_i \qquad (9.3)$$

With these two control equations on hand, one can proceed with the FIS model for the dam according to the following steps:

1. The state variables (h and v) fuzzy subsets are defined in Table 9.8 and Table 9.9, respectively, in a numerical manner with their graphs in Figures 9.13 and 9.14, respectively.
2. The control output MDs (Table 9.10) and runoff MFs (Figure 9.15) are also presented.
3. In light of the aforementioned fuzzy subsets for the decision and control variables, the FAM between these variables is indicated in Table 9.11.
4. Initial conditions will be defined and the calculations performed for four iterations. The initial conditions will be considered at 1000-m elevation with a negative downward volume of -20 m^3. The dynamic recursive equations between the variables were already presented above.

With the initial $h_0 = 1000$ m and $v_0 = -20$ m^3, "L" and "M" fuzzy subsets in the elevation variable are triggered with MDs 1.0 and 0.6, respectively. Volume fuzzy subset "DL" is triggered with the MD of 1.0. Hence, the rule bases from Table 9.11 can be written as:

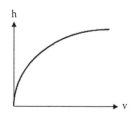

FIGURE 9.12 Volume elevation relation.

TABLE 9.8
Elevation MDs

Elevation (m)	0	100	200	300	400	500	600	700	800	900	1000
"Large," "L"	0	0	0	0	0	0	0.2	0.4	0.6	0.8	1.0
"Medium," "M"	0	0	0	0	0.2	0.4	0.6	0.8	1	0.8	0.6
"Small," "S"	0.4	0.6	0.8	1.0	0.8	0.6	0.4	0.2	0	0	0
"Near zero," "NZ"	1.0	0.8	0.6	0.4	0.2	0	0	0	0	0	0

TABLE 9.9
Volume Increment MDs

Volume Increment (m^3)	−30	−25	−20	−15	−10	−5	0	5	10	15	20	25	30
"Up large," "UL"	0	0	0	0	0	0	0	0	0	0.5	1.0	1.0	1.0
"Up small," "US"	0	0	0	0	0	0	0	0.5	1.0	0.5	0	0	0
"Zero," "Z"	0	0	0	0	0	0.5	1.0	0.5	0	0	0	0	0
"Down small," "DS"	0	0	0	0.5	1.0	0.5	0	0	0	0	0	0	0
"Down large," "DL"	1.0	1.0	1.0	0.5	0	0	0	0	0	0	0	0	0

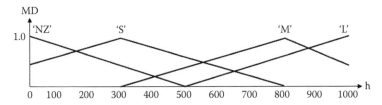

FIGURE 9.13 Elevation MFs.

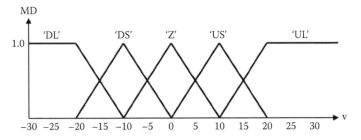

FIGURE 9.14 Volume MFs.

TABLE 9.10
Runoff MDs

Runoff (m³)	−30	−25	−20	−15	−10	−5	0	5	10	15	20	25	30
"Up large," "UL"	0	0	0	0	0	0	0	0	0	0.5	1.0	1.0	1.0
"Up small," "US"	0	0	0	0	0	0	0	0.5	1.0	0.5	0	0	0
"Zero," "Z"	0	0	0	0	0	0.5	1.0	0.5	0	0	0	0	0
"Down small," "DS"	0	0	0	0.5	1.0	0.5	0	0	0	0	0	0	0
"Down large," "DL"	1.0	1.0	1.0	0.5	0	0	0	0	0	0	0	0	0

R1: "IF elevation is 'L' AND volume is 'DL' THEN runoff is 'Z.'"

R2: "IF elevation is 'M' AND volume is 'DL' THEN runoff is 'US.'"

The "minimization" operator on the antecedent parts yields the following statements with the consequent part:

$$\min(1.0,\ 1.0) = 1.0\text{"Z"}$$

$$\min(0.6,\ 1.0) = 0.6\text{"US"}$$

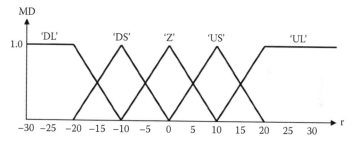

FIGURE 9.15 Runoff MFs.

TABLE 9.11
FAM of Dam Filling

h \ v	"DL"	"DS"	"Z"	"US"	"UL"
"L"	"Z"	"DS"	"DL"	"DL"	"DL"
"M"	"US"	"Z"	"DS"	"DL"	"DL"
"S"	"US"	"US	"Z"	"DS"	"DL"
"NZ"	"UL"	"UL"	"Z"	"DS"	"DS"

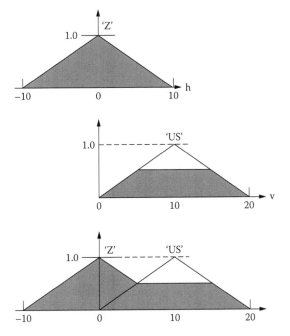

FIGURE 9.16 Defuzzification for iteration 1.

The consequent part of each rule is given graphically in Figure 9.16. The defuzzification with the centroid method leads to $r_0 = 5.5$ m³. The results for this iteration are presented in Figure 9.16. The new state of variables can be calculated by the state equations as:

$$h_1 = h_0 + v_0 = 1000 + (-20) = 980 \text{ m}$$

$$v_1 = v_0 + r_0 = -20 + 5.8 = 14.2 \text{ m}^3$$

With these new initial values for the second iteration from the elevation fuzzy subsets in Figure 9.13, the "L" and "M" MFs are triggered with the respective MDs of 0.96 and 0.64. The new volume value triggers the "DS" and "DL" fuzzy subsets with 0.58 and 0.42 MDs, respectively. These triggers lead to the following fuzzy rule bases from Table 9.11:

R1: "IF elevation 'L' AND volume is 'DS' THEN runoff is 'DS.'"

R2: "IF elevation 'L' AND volume is 'DL' THEN runoff is 'Z.'"

R3: "IF elevation 'M' AND volume is 'DS' THEN runoff is 'Z.'"

R4: "IF elevation 'M' AND volume is 'DL' THEN runoff is 'US.'"

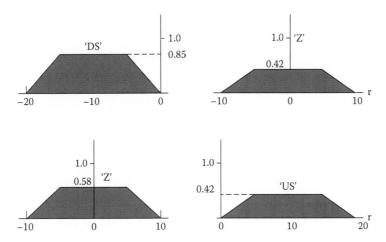

FIGURE 9.17 Defuzzification for iteration 2.

With respective "minimization," the consequent parts take the following results, respectively:

$$\min(0.96, 0.58) = 0.58\text{"DS"}$$

$$\min(0.96, 0.42) = 0.42\text{"Z"}$$

$$\min(0.64, 0.58) = 0.58\text{"Z"}$$

$$\min(0.64, 0.42) = 0.42\text{"US"}$$

The defuzzified runoff is found as $r_1 = -0.5$ m³ and the results are shown in Figure 9.17.

These new values will lead to new initial decision and control values for the next stage as:

$$h_2 = h_1 + v_1 = 980 + (-14.2) = 965.8 \text{ m}$$

$$v_2 = v_1 + r_1 = -14.2 + (-0.5) = -14.7 \text{ m}^3$$

These values trigger from the elevation set of MFs "L" and "M" fuzzy subsets with 0.93 and 0.67 MDs, respectively. The volume triggers are at the "DL" and "DS" fuzzy subsets with 0.43 and 0.57 MDs, respectively. These values lead to the following rule bases from Table 9.11:

R1: "IF elevation is 'L' AND volume is 'DL' THEN runoff is 'Z.'"

R2: "IF elevation is 'L' AND volume is 'DS' THEN runoff is 'DS.'"

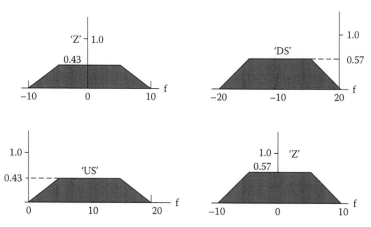

FIGURE 9.18 Defuzzification for iteration 3.

R3: "IF elevation is 'M' AND volume is 'DL' THEN runoff is 'US.'"

R4: "IF elevation is 'M' AND volume is 'DS' THEN runoff is 'Z.'"

Their "minimization" operation at the antecedent parts leads to the consequent parts as:

$$\min(0.93, 0.43) = 0.43\text{"Z"}$$

$$\min(0.93, 0.57) = 0.57\text{"DS"}$$

$$\min(0.67, 0.43) = 0.43\text{"US"}$$

$$\min(0.67, 0.57) = 0.57\text{"Z"}$$

respectively. The defuzzification procedure with centroid crisp runoff yields $r_2 = -0.4$ and the results are presented graphically in Figure 9.18. Again, the computation of the state and control values for the next iteration leads to:

$$h_3 = h_2 + v_2 = 965.8 + (-14.2) = 951.1 \text{ m}$$

$$v_3 = v_2 + r_2 = -14.2 + (-0.4) = -15.1 \text{ m}^3$$

which leads to triggers of "L" and "M" fuzzy subsets with 0.9 and 0.7 MDs, respectively. "DS" and "DL" fuzzy subsets are triggered in the volume domain with MDs of 0.49 and 0.51, respectively. The fuzzy rule bases for these triggers can be appreciated from Table 9.11 as:

R1: "IF elevation is 'L' AND volume is 'DS' THEN runoff is 'DS.'"

R2: "IF elevation is 'L' AND volume is 'DL' THEN runoff is 'Z.'"

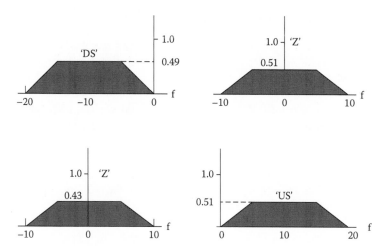

FIGURE 9.19 Defuzzification for iteration 4.

R3: "IF elevation is 'M' AND volume is 'DS' THEN runoff is 'Z.'"

R4: "IF elevation is 'M' AND volume is 'DL' THEN runoff is 'US.'"

With the "minimum" consequent part operations as

$$\min(0.9, 0.49) = 0.49\text{"DS"}$$

$$\min(0.9, 0.51) = 0.51\text{"Z"}$$

$$\min(0.7, 0.49) = 0.49\text{"Z"}$$

$$\min(0.7, 0.51) = 0.51\text{"US"}$$

the defuzzification procedure yields $r_3 = 0.3$ m^3 (see Figure 9.19).
The new iterations have the height and velocity values as follows:

$$h_4 = h_3 + v_3 = 951.1 + (-15.1) = 936.0 \text{ m}$$

$$v_4 = v_3 + r_3 = -15.1 + 0.3 = -14.8 \text{ m}^3$$

The summary of all the iterations is shown in Table 9.12 collectively.

TABLE 9.12
Summary Results of Dam Filling

Iteration	0	1	2	3	4
Elevation (m)	1000/0	980.0	965.8	951.1	926.0
Volume (m³)	−20	−14.2	−14.7	−15.1	−14.8
Runoff (m³)	5.8	0.5	−0.4	0.3	

9.5 CRISP AND FUZZY DYNAMIC PROGRAMMING

Suharyato and Gaulter (2003) presented an application of fuzzy theory as a strategy to produce an operation policy of a single reservoir designed under imperfect data conditions, which is a common feature in developing countries. Their approach considers imperfection (fuzziness) in streamflow data and water demand values. The technique was applied to the Wadaslintang Reservoir in Central Java, Indonesia. It was demonstrated that the incorporation of fuzzy theory in dynamic programming results in more straightforward analysis than the conventional method and yields better values of system performance measures.

Streamflow data of the main river and its tributaries at a reservoir site are vital in the analysis of water balance and operation planning of the reservoir. Unfortunately, reliable and long streamflow records are usually scarce, especially in developing countries. Most often, data is available or only streamflow data of the main river is available, while the data of its tributaries and the surrounding streams is either not available or is only available as qualitative information.

In this situation of imperfect streamflow data condition, one common strategy of solving the crisis of inadequate data is to generate streamflow data from either physiographic information, spatial interpolation, or spatial disaggregation based on the reference data from gauged neighboring basins with similar hydrologic conditions. Because reliable streamflow data in developing countries with low accessibility to remote areas is very scarce—even from surrounding basins—this generation process itself is of questionable nature. The primary concern in this situation is the precision uncertainty of the reference data of the measured rivers, which is passed through and may be exaggerated in the generated data. The resulting analysis, therefore, inherits a bigger imprecision uncertainty, if not misdirected efforts.

The emergence of fuzzy theory (Zadeh, 1965) with its wide prospects of application in many fields bogged with imprecision and qualitative information (Esogbue et al., 1992) offers some hope in properly addressing the problem of imperfect streamflow data condition.

The envisaged scenario in the sequel is reservoir operation analysis during the planning stage with an unreliable and/or short record of streamflow data at the main river and the surrounding streams. As a consequence, the water availability from the reservoir, which can be obtained from water balance analysis, would tend to result in qualitative information as well. Therefore, a rigorous statistical analysis under the condition of imperfect data may not necessarily be appropriate. The inherent imprecision of the data dictates the need to employ fuzzy theory, which is capable of recognizing imprecision in the data and making inferences from it.

In general, a methodological scheme as shown in Figure 9.20 is followed. It is assumed that streamflow data into the reservoir and water demands from the reservoir are adequately represented as fuzzy numbers. This assumption is justifiable considering that imperfect streamflow data and qualitative water demand data constitute the only information commonly found during the planning stage in developing countries.

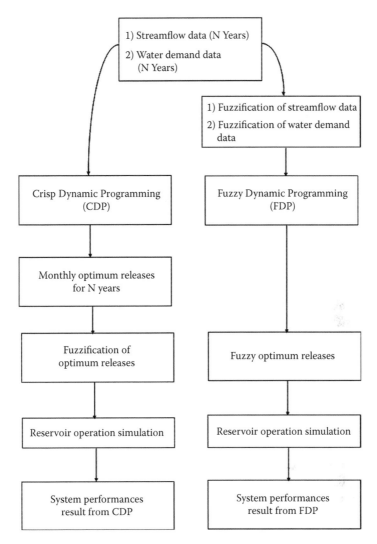

FIGURE 9.20 Reservoir operation by crisp and fuzzy dynamic programming.

One typical recursive equation of dynamic programming is given by (Karamouz and Houck (1982) as:

$$f_t^n(j) = \min_j \left\{ f_{t+1}^{n-j}(i) + (R_{i,j,t} - T_t)^2 \right\} \tag{9.4}$$

where T_t is the water demand and $f_t^n(j)$ is the optimum objective function at the beginning of period t with n periods of the optimization stages remaining if the current period storage state is in class j (Figure 9.21). This variable is also known

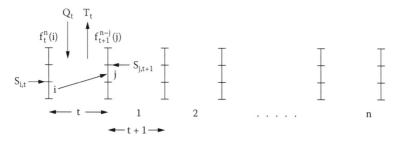

FIGURE 9.21 Storage system evolution by time.

as long-term return, while the $(R_{klt} - T_t)^2$ term is called the *short-term* or *immediate return*. The continuity equation is given by (Loucks et al., 1981):

$$R_{i,j,t} = S_{i,t} + Q_t - L_{i,j,t} - S_{j,t+1} \tag{9.5}$$

where $R_{i,j,t}$ is the release at month t if the current period storage volume is in class i and the next period storage volume is in class j, Q_t is incoming inflow into the reservoir at month t, $S_{i,k,t}$ is the current month (t) storage volume that is in class i, and $S_{j,t+1}$ is the next month ($t + 1$) storage volume that is in class j. $L_{i,j,t}$ is the assumed possible evaporation loss if the current month storage state is at i and the next month storage state is at j. However, these losses are not being considered in order not to mask the primary scheme with incidental complexity.

The main difference between fuzzy dynamic programming and its crisp counterpart lies in storage classes, streamflow, and water demands as fuzzy numbers. As a consequence of fuzzy storage classes and streamflow data, the resulting $R_{i,j,t}$ in Equation (9.4) is also a fuzzy number. In addition, the short-term return in Equation (9.5) is to be reformulated as the square of fuzzy deviation. One index that can be used to measure fuzzy deviation is a dis-resemblance index as shown by Kaufmann and Gupta (1985). This index is denoted by d("A", "B") and has a value in the range between 0 and 1. This index is defined as:

$$\delta('A', 'B') = \frac{A_A + A_B + 2A_s - A_I}{2(R_L - R_U)} \tag{9.6}$$

where A_A and A_B are the area under the MFs for fuzzy numbers "A" and "B," respectively; A_s is the area of separation between the MFs for the two fuzzy numbers "A" and "B"; A_I is the area of intersection between the MFs for the two fuzzy numbers "A" and "B"; and finally, R_L and R_U are the lower and upper boundaries of the universe, respectively.

The monthly release rules derived from crisp dynamic programming (CDP) formulation are denoted by "R_t" $= [\alpha_t, R_t^{ave}, \beta_t]$, where $\alpha_t = R_t^{ave} - R_t^{min}, R_t^{ave}$ and $\beta_t = R_t^{max} - R_t^{ave}$ are the left spread, the median, and the right spread of the fuzzy release "R_t", respectively. R_t^{min} and R_t^{max} are the minimum and maximum optimum releases at month t, respectively, obtained from CDP formulation.

A simulation run of a single reservoir was conducted for a period of 25 years, which is equal to the length of time of reference data generally used in small reservoir studies. In the operation phase, the reservoir standard operating rule is employed:

$$R_t = \begin{cases} S_t + Q_t - S_{max} & \text{if} \quad S_t + Q_t - T_t > S_{max} \\ T_t & \text{if} \quad S_{min} \leq S_t + Q_t - T_t \leq S_{max} \\ Q_t & \text{if} \quad S_t + Q_t - T_t < S_{min} \end{cases} \tag{9.7}$$

where S_t is the simulated storage volume at month t, and S_{max} and S_{min} are the maximum and the minimum storages, respectively. Whenever the release falls outside the range of $\left[R_t^{left}, R_t^{right} \right]_\alpha$ associated with the α-cut of the corresponding monthly fuzzy release rule, it is adjusted according to:

$$R_t' = \begin{cases} R_t^{right} & \text{if} \quad R > R_t^{right} \\ R_t & \text{if} \quad R_t^{left} \leq R_t \leq R_t^{right} \\ 0 & \text{if} \quad R_t < R_t^{left} \end{cases} \tag{9.8}$$

The storage volume after releasing $R_{t'}$ is calculated from the continuity Equation (9.5). The performance of the fuzzy release rules is defined by:

$$\gamma_\alpha = 100 \frac{N_{in}}{N_{sim}} \tag{9.9}$$

and

$$\theta_\alpha = 100 \frac{\displaystyle\sum_{i=1}^{N_{in}} \mu_{R_t}(R_{t'})}{N_{in}} \tag{9.10}$$

where γ_α (in percent) is the reliability of fuzzy release rules at α-cut level θ_α (in percent) is the average quality of resemblance of the $R_{t'}$ with the fuzzy release rule at α-cut level, N_{in} is the frequency of $R_{t'}$ in the range of $\left[R_t^{left}, R_t^{right} \right]_\alpha$, N_{sim} is the period of simulation, and $\mu_{R_t}(R_{t'})$ is the membership level of the fuzzy release rule of month t at the value of $R_{t'}$. To evaluate the consequence of restricting the release to be in the range of $\left[R_t^{left}, R_t^{right} \right]_\alpha$, the reliability of the system is also measured. The system reliability, denoted as η, is calculated as:

$$\eta = 100(1.0 - F_f) \tag{9.11}$$

where F_f is the frequency of failure. The reservoir policy is considered to have failed if either

1. The release $R_{t'}$ is less than the water demand T_t, or
2. The storage volume S_t is less than the minimum storage S_{min}

Wadaslintang reservoir is located on the Bedegolan River in Central Java, Indonesia. The reservoir is mainly used to supply the water demands of the surrounding irrigation area. Its purposes include hydro power generation of 2×8 MW capacity, low flow augmentation, and water supply to meet domestic, municipal, and industrial demands.

During the planning stage in 1978, there was not sufficient streamflow data of the Bedegolan River and the surrounding streams within the irrigation system. To conduct water balance analysis and operation planning, the consultant had to generate streamflow data with reference to the Jragung/Tuntang watershed (ECI, 1978). Thus, the generated data and the resulting water demands are referred to as historical data, and will be used to derive fuzzy streamflow data, fuzzy water demands, and stochastically generated data of streamflow and water demands.

A comparison of the execution time between CDP and fuzzy dynamic programming (FDP) programs for different values of the number of storage (NS) classes is shown in Figure 9.22. It is shown that the FDP program requires much less execution time. This situation can be explained by the fact that the optimization in CDP program is conducted for the entire time horizon of N years, while in FDP program, it is conducted for 1 year only. It is later shown, however, that this straightforward solution of the FDP program does not hinder a better analysis.

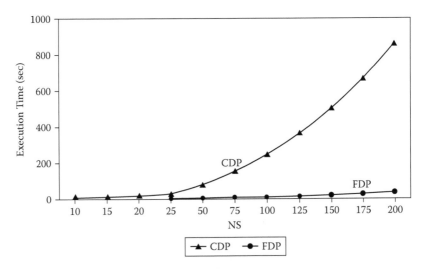

FIGURE 9.22 Execution time of CDP and FDP programs.

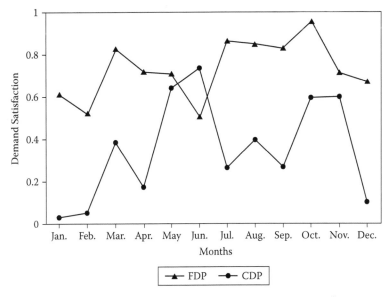

FIGURE 9.23 Satisfaction level of fuzzy release to average monthly demand.

To avoid the "trapping" effect of the storage level as has been shown, for example, in Goulter and Tai (1985) and the "overlooking" of small water demands, the number of storage classes adopted in further analysis is 200 intervals.

A straightforward comparison of the fuzzy release rules generated by CDP and FDP programs is shown in Figure 9.23. The figure shows the satisfaction level of the fuzzy release rules at the average values of monthly water demands. It clearly shows that the satisfaction levels of the fuzzy release rules generated by the FDP program are higher than those from the CDP program most of the time. This result shows that the adaptation of the fuzzy release rule tends to result in a higher possibility of supplying the demand, in this case, at its average value.

The summary of simulated performance measures is shown in Figure 9.24, which shows the measure of release rule reliability at the α-cut level, the quality of release resemblance to the fuzzy release rule at α-cut level, and the reservoir reliability as a consequence of confining the release in the range of α-cut level of fuzzy release rule. In general, the three measures resulting from the FDP program always show better values compared to those from the CDP program at all levels of α-cut.

It is demonstrated that the application of fuzzy theory to the condition of imperfect data is beneficial and leads to more straightforward analysis than conventional methods, yet results in better system performances. Fuzzy theory holds great promise in the field of reservoir operation studies because of the large amount of uncertainty present in the system.

Since conditional statements are used to make complete sentences, "IF . . . THEN . . ." rules are used to make something useful in FL.

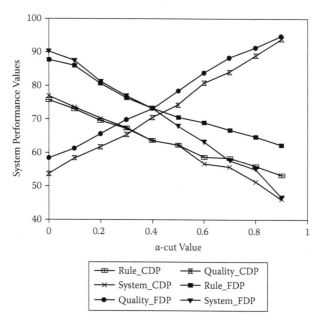

FIGURE 9.24 System performances of Wadaslintang reservoir.

9.6 MULTIPLE RESERVOIR OPERATION RULE

A Mamdani FIS model is constructed to derive joint operation rules for water supply reservoirs. The case study of Yıldız (Istranca) Creek reservoirs, together with Terkos Lake on the European side of Istanbul, illustrates the proposed methodology (see Figure 8.35 in Chapter 8). As the first stage, an additional water supply is brought to Istanbul from Yıldız (Istranca) mountain creeks through three small reservoirs (Buyukdere, Kuzuluder, and Duzdere). The water collected in these reservoirs is carried to Terkos Lake, then onward to Istanbul through pipelines. It is therefore necessary to develop an operation program for optimum management of the water supply from these three creek reservoirs, first to Terkos Lake and then directly to water treatment plants in Istanbul. Although such an optimization and management report was prepared by Şen (1996), the methodology used was completely deterministic, with no consideration for uncertainties in detail except in the simple statistical sense. However, with increasing additional reservoirs in the Yıldız Mountains, it became necessary to develop an operation program that would account for imprecision and vagueness, which are ingredients of the input and many other data.

Operation rules are generated on the basis of a sustainable water supply for meeting the municipal and industrial demands as well as environmental criteria such as water quality for fish and wildlife preservation, recreational needs, and downstream flow regulation. The proposed fuzzy rule-based model operates on an "IF . . . THEN . . ." principle. In the premise part the reservoir storage levels, estimated inflows, evaporation losses, and probable precipitation amounts are employed, each with convenient fuzzy subsets. The release from each reservoir is adopted in the form of another fuzzy subsets as the consequent part of the fuzzy implication.

Even a single real-world reservoir operation model can be very complex and therefore requires a different set of rules for its joint management with other nearby reservoirs. This is due to the fact that input variables are full of imprecision and vagueness, and output decision variables must satisfy the required set of restrictions without violating the physical constraints of the system concerned. Among the physical constraints are the reservoir capacities, maximum and minimum water levels, pipe capacities, and the demand for water supply. In addition, the input variables are incorporated with uncertainty, and therefore their future projections based on the past records are established through probabilistic, statistical, and stochastic approaches. In fact, stochastic reservoir control, operation, and maintenance procedures have been applied to interconnected reservoir systems for the past 50 years. Most often, input variables such as the runoff, direct rainfall, snowfall, evaporation, and infiltration have been modeled by various levels of complex stochastic processes, starting with simple Markov, autoregressive, and moving average processes and their combinations in the form of autoregressive integrated moving average (ARIMA) processes. All such processes treat numerical uncertainties in the input data in terms of probability distribution functions (PDFs), which are the prerequisites for any stochastic modeling.

It is true in practical applications, especially at the level of operators, that the input variables might not incorporate numerical imprecision but rather linguistic imprecision such as "high flow," "flood," "severe runoff," "low evaporation," "dry period," and like. These linguistic expressions cannot be modeled by any means except FL approaches, the basis of which was first proposed by Zadeh (1965). Hence, FL becomes the most appropriate tool in handling such linguistic imprecision because the uncertainty does not lie only in the value of the input variable, but also in the extent to which these variables belong to a given imprecision set. In fact, a given value of inflow may belong simultaneously to more than one fuzzy set with different extents. Fuzzy sets and logic achieve the treatment of such linguistic imprecision in variables that cannot be accounted for by any stochastic process. It is possible that the PDF of the variable might not be available. Such a situation hinders the use of any stochastic process application in uncertainty modeling.

FIS modeling of a set of reservoirs in an interconnected manner is a relatively simple approach, where the rules operate on "IF . . . THEN . . ." type of logical principles. IF is a set of fuzzy explanatory variables or premises, such as reservoir water level inflows, forecast demand stages, and time of year (Figure 9.21). On the other hand, THEN includes a fuzzy consequence, such as the actual releases from the reservoirs. By considering all possible "IF . . . THEN . . ." statements concerning the single or joint operation of reservoirs, they are then interconnected by logical "AND" or "OR" connectivity, depending on expert views and the experience of the reservoir operators. The rules given by the fuzzy rule bases can be executed according to the evaluation concepts of engineering sustainability and risk.

The general situation and location of the three reservoirs with respect to Terkos Lake are shown in Figure 9.25, together with water resource quantities of the three creeks in Table 9.13. In Figure 9.13, r, P, E, and O indicate the runoff, precipitation, evaporation, and output of the respective dams with corresponding subscripts.

For the FL operation of Terkos Lake with these three reservoirs, first it is necessary decide on the MFs of the various input, output, and reservoir storage variables

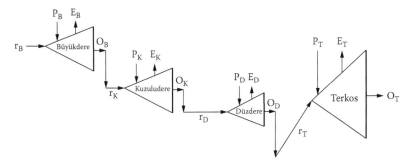

FIGURE 9.25 Terkos Lake and the three Yildiz mountain reservoirs.

TABLE 9.13
Some Features of the Yildiz Mountain Reservoirs

Creek Name	Catchment Area (km²)	Monthly Flow (×10⁶ m³)		Volume (×10⁶ m³)	
		Average	Maximum	Active	Yield
Düzdere	10.5	1.2	3/7	0.12	4.5
Kuzuludere	31.0	2.59	5.45	1.6	13.5
Büyükdere	50.0	2.76	6.10	2.6	28.4
			Total	4.32	46.4

according to the data at hand. Herein, independent of the data, a general partition of the relevant variables into fuzzy subsets and their MFs in the form of triangular series is presented. For this purpose, the storage (i.e., the volume partition of the reservoirs) is based on the maximum (V_M) and minimum (V_m) volumes designed for each reservoir. Five different but concurrent stages of the storage are considered: "empty" ("E"), "almost empty" ("AE"), "medium" ("M"), "almost full" ("AF"), and "full" ("F"). Each stage is represented by triangular MFs as in Figure 9.26 wherein the subdivisions are also shown. It is obvious that all the MFs depend on V_M and V_m values only. The input variables (precipitation, runoff) and output variable from the reservoir surface area as evaporation are all represented by a standard MF in a series of triangles that is defined this time by considering the arithmetic average (X) and the standard deviation (σ_X) of these variables. Figure 9.27 shows the MFs with "very

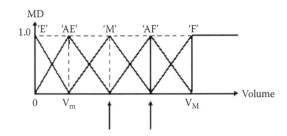

FIGURE 9.26 Volume MFs.

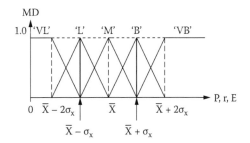

FIGURE 9.27 Precipitation, runoff, and evaporation MFs.

Little" ("VL"), "little" ("L"), "medium" ("M"), "big" ("B"), and "very Big" ("VB") linguistic variables in terms of five fuzzy functions. As far as the decision variables are concerned, the output (O) from each reservoir is divided into six fuzzy subsets— "very little" ("VL"), "little" ("L"), "medium" ("M"), "big" ("B"), "very big" ("VB"), and "extremely big" ("EB")—as shown in Figure 9.28. In the same figure on the horizontal axis, the maximum pump capacity is shown as C.

For practical applications it is possible to apply fuzzy dynamic program as explained in the previous section. The classical operation and management of the aforementioned reservoirs with Terkos Lake were presented by S ̧en (1996) and Kadıog ̆lu and S ̧en (2001) on the basis of the dynamic programming technique, where no elements of uncertainty or imprecision in the variables were considered. To appreciate the joint operation, it is first illuminative to consider the interconnected system of Terkos Lake with Düzdere reservoir. The output from the Düzdere reservoir will depend on the following three restrictions through the fuzzy implications where the premises are Terkos Lake volume stage restrictions. The "Crisp Constraint Stage" has the following steps:

1. IF V_{TE} is greater than O_D THEN O_D is one of the inputs into the lake.
2. IF V_{TE} is less than O_D THEN V_{TE} is a transmission input into the lake.
3. IF V_{TE} is equal to zero THEN Terkos Lake should be operated separately.

In these statements, V_{TE} is the Terkos Lake empty volume (i.e., the minimum volume, V_m) and O_D is the Düzdere output. Because there is constant demand is for

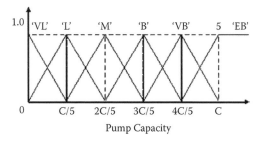

FIGURE 9.28 Demand MFs.

Istanbul water supply from Terkos Lake, it will be operated according to the following steps:

1. Decide about the initial volume (V_I), area (A_I), and constant demand (D).
   a. Enter precipitation P_i, runoff r_i, and evaporation E_i values into fuzzy
      rule bases in Figure 9.26 through "Fuzzy Internal Stage Machine" find
      the triggering consequent rules and then come out with a fuzzy output
      volume, V_f from Figure 9.25.
   b. Find the final volume V_F as $(V_I + V_f)$,
   c. Control the final volume according to the following crisp statements
      i.   IF V_F is greater than V_M THEN V_F is equal to V_M and the spill is $s =$
           $V_F - V_M$,
      ii. IF V_F is less than V_M THEN V_F is equal to V_m and the deficit is $d =$
           $V_m - V_F$,
      iii. Otherwise, V_F remains as it is.
   d. Find output value from execution of fuzzy final stage,
   e. IF final operation period is not reached then repeat each step from (A)
      onwards,
   f. Otherwise end.
2. Find the final volume as $V_F = (V_I + V_f - D)$.
3. Execute the 'Crisp Constraint Stage'.
4. Execute (v).
5. Stop.

9.7 LAKE LEVEL ESTIMATION

The main purpose of this section is to develop an estimation procedure independent
of the crisp autocorrelation concept and accordingly restrictive assumptions of linearity, normality (Gaussian distribution), homoscedasticity (variance constancy), and
stationary. FIS is suggested and then used to predict monthly lake level fluctuations.
This methodology is capable of depicting the non-linearity, non-normality, and non-
stationary features in lake level fluctuations. Given a sequence of lake levels as H_1,
$H_2, \ldots, H_n$ where n is the number of measurements, it is possible to develop for the
next, H_i, lake level an estimation from two previous levels, H_{i-1} and H_{i-2}. For this purpose, a Sugeno FIS is suggested as in Figure 9.29 and all previous steps are included
in this schematic model.

Inputs and their MFs appear to the left of the FIS structural characteristics, while
outputs and their MFs appear on the right. The fuzzy subsets of input are presented
in Figure 9.29. Outputs of each rule are given as a linear combination of inputs:

$$f_r(H_{i-1}, H_{i-2}) = a + bH_{i-1} + cH_{i-2} \qquad (9.12)$$

Lake Van water level records are used for the implementation of the Sugeno FIS
methodology so as to depict the common behavior of three variables, which are

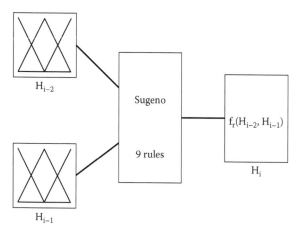

FIGURE 9.29 TS fuzzy model input-output diagram.

taken consequently from the historical time series data. The first two variables represent the two past lake levels and the third one indicates present lake levels. Hence, the model has three parts, namely, observations (recorded time series) as input, fuzzy model as response, and the output as estimations. It is possible to consider lags between the successive data at one, two, or three lags. Such an approach is very similar to a second-order Markov process, which can be expressed as:

$$H_i = \alpha H_{i-1} + \beta H_{i-2} + \varepsilon_i \tag{9.13}$$

where H_i, H_{i-1}, and H_{i-2} are the three consecutive lake levels; α and β are model parameters; and ε_i is the error term with zero mean. The application of such a model requires the parameter estimations from the available data prior to any estimation procedure. Furthermore, its application is possible in light of a set of assumptions that includes linearity, normality (Gaussian distribution of the residuals, i.e., ε_i's), variance constancy, ergodicity, and independence of residuals. The Sugeno FIS model replaces Equation (9.13) without any restriction. It is considered an appearance of the natural relationship between three consecutive time values of the same variable.

To apply the Sugeno FIS approach, it is necessary to divide the data into training and testing parts. Herein, the 24 months (2 years, 1993–1994) are left for the test (estimation), whereas all other data values are employed for training. Altunkaynak et al. (2003) identified suitable models and provided estimates for lake level fluctuations and their parameters for trend, periodic, and stochastic parts. A second-order Markov model is found suitable for the stochastic part. Later, they established the triple diagram model of lake levels, which is suggested as a replacement of the second-order Markov process. It is not necessary to use first- and second-order autocorrelation coefficients for taking into account persistence.

The system with three MFs for each input and nine rules is determined as the most suitable one after the training process. The input MFs are shown in Figure 9.30. Triangular MFs that provide smaller relative error among the other types are selected

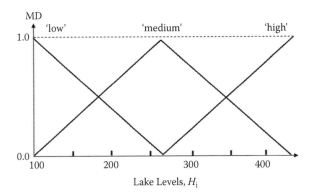

FIGURE 9.30 Lake level MFs.

for modeling. The last column of Table 9.14 shows the output model parameters of the output functions as in Equation (9.12).

There are nine fuzzy rules for the verbal expression of the relationship between three successive lake levels. An example for the inference algorithm is given in Table 9.15. The second and third columns indicate the fuzzy sets of the input variables. Four rules—namely, R5, R6, R8, and R9—are triggered and thus the output value is the weighted average of the four output functions according to Equation (7.8). The fifth and sixth columns are for the MDs. Each rule contributes to the result with different weights. There are two ways to determine these weights. Herein, the product of the antecedent MDs is used; it gives smaller relative error than "min-max" FIS (Chapter 7).

The prediction results are shown in Table 9.16 with corresponding relative error amounts. Individual errors are slightly greater than 10 percent, but the overall

TABLE 9.14

Sugeno FIS Model Parameters

Rule No.	Input		Output		
	H_{i-2}	H_{i-1}	H_i		
			a	b	c
1	"low"	"low"	51.50	−310.12	310.60
2	"low"	"medium"	175.27	−815.18	310.69
3	"low"	"high"	−114.81	−1369.67	393.78
4	"medium"	"low"	71.18	−311.40	816.65
5	"medium"	"medium"	106.19	−814.84	815.44
6	"medium"	"high"	−96.28	−1320.12	816.87
7	"high"	"low"	−51.61	2824.50	−1975.80
8	"high"	"medium"	−89.00	−815.03	1320.88
9	"high"	"high"	71.21	−1319.26	1320.08

TABLE 9.15
Rule Triggering

Rule No.	$H_{i-2} = 335$	$H_{i-1} = 340$	f_i	MD		α_i	$\alpha_i f_i$
				H_{i-2}	H_{i-1}		
R5	"M"	"M"	4386.60	0.5555	0.5276	0.2931	1285.6519
R6	"M"	"H"	−164599.38	0.5555	0.4724	0.2624	−43193.1999
R8	"H"	"M"	274380.17	0.4384	0.5276	0.2313	40703.0349
R9	"H"	"H'	175972.74	0.4384	0.4724	0.2232	1550.2836
			Total			1.0100	345.7706

TABLE 9.16
Lag-One Lake Levels Prediction (cm)

H_{i-2}	H_{i-1}	H_i	Fuzzy Prediction	Relative Error (%)
274	285	293	289.51	1.19
285	293	305	293.95	3.62
293	305	332	308.54	7.07
305	332	354	351.57	0.69
332	354	355	374.37	5.17
354	355	354	357.38	0.95
355	354	343	354.57	3.26
354	343	333	333.71	0.21
343	333	335	322.69	3.67
333	335	340	334.49	1.62
335	340	347	342.36	1.34
340	347	350	352.91	0.82
347	350	356	353.16	0.80
350	356	370	362.73	1.97
356	370	392	385.76	1.59
370	392	396	415.88	4.78
392	396	393	401.76	2.18
396	393	384	393.10	2.32
393	384	372	380.15	2.14
384	372	364	365.81	0.50
372	364	364	360.32	1.01
364	364	367	366.85	0.04
364	367	373	372.72	0.07
367	373	377	381.84	1.27
Average				**2.01**

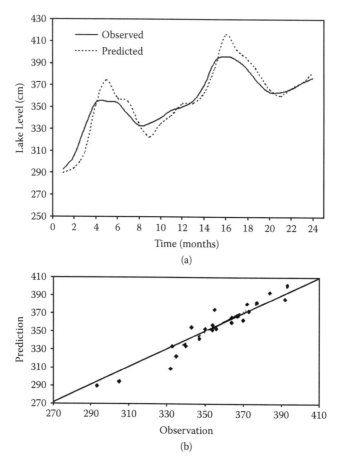

FIGURE 9.31 For the test data at lag-one: (a) lake levels time series and (b) model verification.

prediction relative error percentage is about 2.01 percent. Figure 9.31a represents lake level observations and prediction fluctuations whereas Figure 9.31b indicates the observed and estimated H_i values. It is obvious that they follow each other very closely and, on average, the observed and estimated lake level series have almost the same statistical parameters.

The Sugeno FIS model even depicts the increasing trend, which is not possible directly with the second-order Markov process. During the fuzzy estimation procedure, there is no special treatment of trend, but even so it is modeled successfully. However, in any stochastic or statistical modeling process, it is first necessary to make trend analysis and separate it from the original data. To further show the verification of the Sugeno FIS approach for lake level estimations, the test data in Figure 9.31b is plotted versus the estimations. It is obvious that almost all the points are near 45° lines, and hence the model is not biased. Estimations are successful at low or high values and have a very high correlation with observations.

9.8   TRIPLE DIAGRAMS RULE BASE

Human beings can visualize, at a maximum, three-dimensional variations. The best configuration and interpretation of such variations can be achieved in three-dimensional Cartesian coordinate systems through contour maps. Generally, maps are regarded as the variation of a variable by location variables that are either longitudes and latitudes or eastings and northings (Cressie, 1993; Isaaks and Srivastava, 1989). Hence, it is possible to estimate the concerned (mapped) variable value for a given pair of location variables. Similarly, because one wants to predict the current lake level from previous records, it is suggested that two previous records replace the two location variables. In this manner, it is possible to map the current values of a variable based on two previous values of the same variable. It is also possible to devise maps through Kriging methodology (Matheron, 1965; 1971; Şen, 2009), where the first step in any Kriging methodology prior to mapping is to determine the empirical semivariogram (SV), which guides the theoretical SV model that will be employed in the mapping procedure. The SV indicates the half-squared variation between two locations with distance. For this purpose, using the same lake data as in the previous section, the scatter of SV values versus distance is obtained for lag-one, lag-two, and lag-three (Figure 9.31). To depict the general trend of the scatter diagram, the distance range is divided into nine intervals, and the average of the SV values that fall within each interval is considered the representative SV value within the mid-point distance of that interval, as suggested by Myers et al. (1982). Different theoretical SV models such as linear, power, spherical, and Gaussian types are tried for the best fit, and at the end, the Gaussian SV is observed to have the best match with the experimental SV trend (see Figure 9.32). The Gaussian model is the most suitable one in all lags, and the properties of fitted Gaussian SV model are presented in Table 9.17.

Such a mapping technique is referred to, herein, as triple diagram methodology (TDM). Such maps are based on three consecutive lake levels. TDMs help make interpretations in spite of extremely scattered points. The construction of a TDM

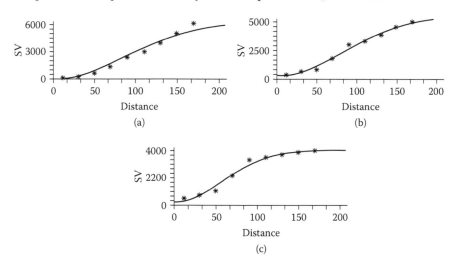

FIGURE 9.32 Empirical and theoretical SVs: (a) lag-one, (b) lag-two, (c) lag-three.

TABLE 9.17

Theoretical Gaussian Semivariogram Parameters

Lag	Nugget (cm²)	Sill (cm²)	Range (cm)	Correlation Coefficient
1	70.0	6250	213.20	0.977
2	270.0	5555.0	586.80	0.990
3	250.0	4327.0	136.0	0.977

requires three variables, two of which are referred to as independent variables (predictors) and constitute the basic scatter diagram. The third variable is the dependent variable, which has its measured values attached to each scatter point. The equal value lines are constructed by the Kriging methodology concept, which is also referred to as geostatistics (Matheron, 1963). Details of this methodology are explained for earth sciences applications by Journel and Huijbregts (1978), Isaaks and Srivastava (1989), Cressie (1993), and Şen (2009).

To apply the triple diagram approach, it is necessary to divide the data into training and testing parts. As in the previous section, the last 24 months (2 years) are left for the test (estimation), whereas all other data values are employed for training, which is the mapping. Maps are prepared according to the Kriging procedure using available software programs.

Prior to any estimation, it is possible to draw the following interpretations from these figures:

1. In the case of lag-one, there is a strong relationship between H_{i-1} and H_{i-2} with increasing contour values of H_i along almost any 45° line (see Figure 9.33a). The 'small' H_i values are concentrated at small H_{i-1} and H_{i-2} values; this implies the clustering of "small" values of the three consecutive lake levels. Similarly, "high" lake level values of the three consecutive levels also constitute "high" values cluster. This means that "small" values follow "small" values and "high" values follow "high" values, which indicates positive correlations (Chapter 5). Local variations in the contour lines appear at either "low" ("high") H_{i-1} or "high" ("low") H_{i-2} values. Consequently, better predictions can be expected within a certain band around the 45° line. It is possible to deduce the following set of logical rules from Figure 9.34:

IF H_{i-1} is "low" AND H_{i-2} is "low" THEN H_i is "low."

IF H_{i-1} is "medium low" AND H_{i-2} is "medium" THEN H_i is "medium."

IF H_{i-1} is "high" AND H_{i-2} is "high" THEN H_i is "high."

These rules can be used for a fuzzy logic inference system, as suggested by Zadeh (1965).

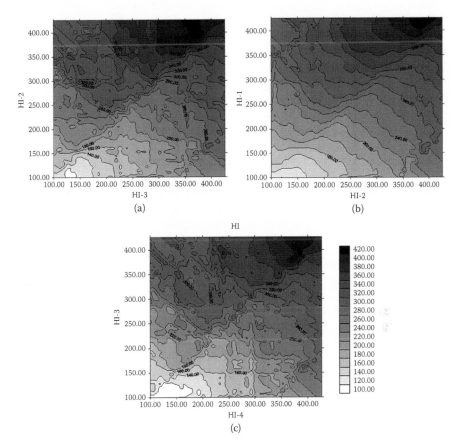

FIGURE 9.33 Lake level maps at (a) lag-one, (b) lag-two, and (c) lag-three.

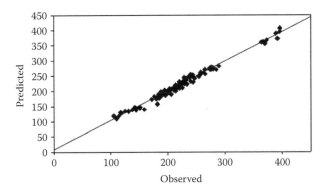

FIGURE 9.34 Lag-one model verification.

2. In Figure 9.32b (lag-two), variations in the contour lines become very distinctive and rather haphazard compared to Figure 9.32a. This implies that with the increment in the lag value, present-time lake level estimation will have more relative error. There is also a distinctive 45° line but with a comparatively narrower band of certainty around it.

3. Finally, in the lag-three case (Figure 9.33c), the contour pattern takes on even more haphazard variation. This implies an increase in the relative error of estimations.

To make estimations for the last 24 months that are not used in the triple diagram constructions in Figure 9.33, it is necessary to enter input values for successive months on the vertical and horizontal axes, respectively. The estimation value of H_i can be either read from the maps approximately or calculated by the Kriging estimation. The estimation results are shown in Table 9.18 for lag-one with corresponding relative error amounts. Individual errors are slightly greater than 10 percent but the overall estimation relative error percentage is approximately 4.83 percent.

TABLE 9.18
Lag-One Lake Level Estimations (cm)

H_{i-2}	H_{i-1}	H_i	Prediction	Relative Error (%)
119	112	110	109.32	0.62
107	111	114	118.40	3.72
125	130	125	134.87	7.32
130	125	118	128.43	8.12
125	118	105	118.60	11.47
120	138	142	145.50	2.41
138	142	138	139.53	1.10
142	138	131	134.16	2.35
138	131	117	131.64	11.12
127	141	151	144.08	4.58
141	151	151	146.85	2.75
151	151	144	144.46	0.32
137	140	144	138.76	3.64
140	144	159	140.22	11.81
144	159	182	157.03	13.72
199	202	195	203.62	4.23
202	195	185	191.33	3.31
195	185	177	182.29	2.90
189	193	202	202.01	0.00
193	202	221	209.94	5.00
202	221	245	229.48	6.34
262	254	244	252.90	3.52
254	244	239	243.16	1.71
244	239	241	231.89	3.78
			Average	**4.83**

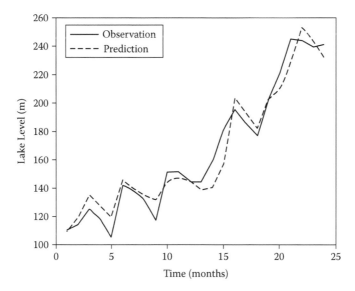

FIGURE 9.35 Observed and predicted lake levels at lag-one for the test data.

Figure 9.35 indicates time series for the observed and estimated H_i values. It is obvious that they follow each other very closely and, on average, the observed and estimated lake level series have almost the same statistical parameters. The TDM model depicts even the increasing trend, which is not possible directly with the stochastic process. During the estimation procedure there is no special treatment of trend but, even so, it is modeled successfully.

It is also possible to look at TDM model performance at lag-two and lag-three. For this purpose, Figure 9.36 suggests that the prediction deviations are larger than in Figure 9.35. However, they still depict the general trend. The prediction results are shown in Table 9.19 and Table 9.20 for lag-two and lag-three, respectively. It is numerically possible to see that an increase in the lag causes an increase in relative error.

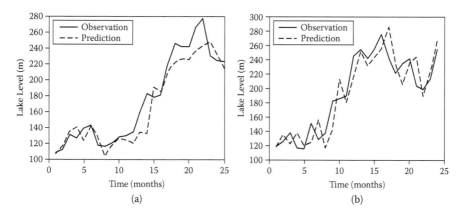

FIGURE 9.36 Observed and predicted lake levels: (a) lag-two and (b) lag-three.

TABLE 9.19
Lag-Two Lake Level Estimations (cm)

H_{i-3}	H_{i-2}	H_i	Estimation	Relative Error (%)
119	112	107	105.87	1.06
112	110	111	115.45	3.85
114	118	130	134.06	3.03
118	125	125	140.47	11.01
106	111	138	121.73	11.79
111	120	142	140.79	0.85
138	131	116	126.71	8.46
131	117	115	101.69	11.57
112	111	119	116.89	1.77
111	115	127	124.38	2.06
151	144	128	123.55	3.47
144	133	133	118.52	10.89
137	140	159	132.72	16.53
140	144	182	131.05	27.99
202	195	177	189.89	6.79
195	185	180	183.49	1.90
189	193	221	209.51	5.20
193	202	245	221.34	9.66
254	244	241	224.79	6.73
244	239	241	224.47	6.86
243	247	266	235.89	11.32
247	255	276	241.96	12.33
268	258	229	246.46	7.09
258	243	223	229.29	2.75
223	221	222	212.15	4.44
			Average	**7.58**

9.9 LOGICAL-CONCEPTUAL MODELS

In current wadi groundwater resources assessments, haphazard and heuristic techniques are employed without logical, rational, or expert views (Şen, 2008). Prior to a wadi groundwater management (WGM) system, it is necessary to obtain conceptual models (logical) in the individual and integrated wadi systems. Herein, conceptual management models are considered that describe essential features of groundwater phenomena and identify the principal processes that take place during a management program. A complete conceptual model provides information regarding the following points:

1. Definition of the phenomenon in terms of features recognizable by observations, analysis, or validated simulations. Herein, the phenomenon includes the groundwater recharge to and exploitation from two wadis near the Red

TABLE 9.20
Lag-Three Lake Level Estimations (cm)

H_{i-4}	H_{i-3}	H_i	Estimation	Relative Error (%)
110	107	118	117.62	0.32
107	111	125	134.91	7.34
104	106	138	121.88	11.68
142	138	116	137.34	15.54
116	115	115	119.66	3.90
115	119	151	124.13	17.80
151	151	128	155.85	17.87
133	128	137	116.73	14.79
137	140	182	142.12	21.91
182	199	185	212.46	12.93
185	177	189	179.75	4.90
189	193	245	214.36	12.51
221	245	254	251.48	0.99
254	244	241	231.53	3.93
241	241	255	244.45	4.14
247	255	275	254.95	7.29
276	275	243	286.02	15.04
258	243	221	231.91	4.70
221	222	234	205.56	12.15
226	234	241	235.28	2.37
244	241	203	243.34	16.58
214	203	198	188.78	4.66
195	198	213	222.06	4.08
213	237	255	268.20	4.92
			Average	**9.26**

Sea, namely, wadis Fatimah and Na'man and their joint integrated operation and management for the best service during a strategic emergency situation. For this purpose, various linguistic and simple logical alternatives are identified.

2. Description of the WGM practices in terms of appearance (various alternatives), size, intensity, and accompanying groundwater conditions.

3. Logical statements about hydro-physical processes, which enable the understanding of the factors that determine the mode and rate of groundwater recharge and consumption with time. This corresponds to the derivation of basic and simple logical rules and regulations about the overall system performance.

4. Specification of the key hydrogeological fields demonstrating the main processes, such as the recharge potential, saturated and unsaturated zone potentials for abstraction, and additional storage of groundwater.

5. Guidance for hydrogeological condition estimations or situations using the diagnostic and prognostic fields that best discriminate between development and non-development guiding to the WGM.

Conceptual models provide decision makers with the following knowledge, which can then be employed in any effective WGM study:

1. *Diagnosis:* This is the preliminary step that helps in understanding the internal and external activities within the entire system under consideration.
2. *Synthesis:* Available verbal and numerical data and information must be synthesized for arriving at preliminary conclusions that help guide toward the optimum pattern of WGM.
3. *Logic:* This provides a "mental picture" of WGM aspects within- and intra-wadi situations.
4. *Isolation:* The basic ingredients such as sub-wadis and WGM variables are isolated from each other so as to assess individual effects.
5. *Main pattern:* This includes the identification of the main and distinctive patterns within the entire complex system.
6. *Numerical evaluation:* Mental and logical aspects are assessed numerically by considering individual and joint WGM operations through a set of rules (rule base).
7. *Numerical projections:* Based on the previous steps, numerical predictions of future situations are achieved for different scenarios.
8. *Numerical products:* Different tools are used for the modification of numerical products.
9. *Simple calculations:* Rather than involved mathematical procedures, it is preferable to construct preliminary logical steps for a simple WGM model.
10. *Data gaps:* Suggested scenarios provide the possibility of filling gaps in the data.

9.9.1 CONCEPTUAL MODEL OF **FATNAM** SYSTEM

The conceptual model of the sub-systems in each wadi (Fatimah and Na'man) is presented in Figure 9.37 for the entire FATNAM management program (Şen, 2005). This is a block diagram that shows the natural groundwater flow from the main wadi branches and along the main channel in each wadi. Note that each sub-unit within the wadi is connected in either a parallel or serial manner. Parallel connection implies that groundwater withdrawal from one of the branches does not affect the other, and hence their contribution to the FATNAM system is independent of each other, which implies "ORing" conjunction. On the other hand, serial connection implies that groundwater withdrawal is interconnected between the components, which requires "ANDing" conjunction. In such a situation, groundwater withdrawal should start at the lower end (downstream) component. This may delay groundwater abstraction because in such a case groundwater transmission from the upper branches to the lower ones takes time. Depending on the aquifer transmissivity value, the time varies and the greater

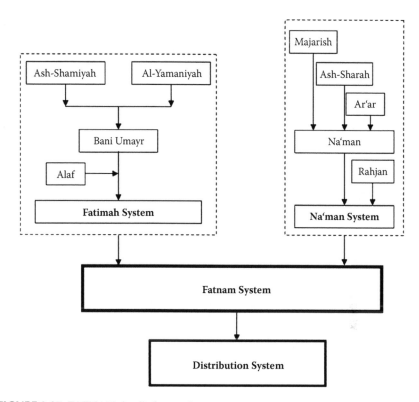

FIGURE 9.37 FATNAM detailed groundwater management system.

the transmissivity, the smaller the groundwater transportation time. In the case of serial connection, the aquifer plays a role similar to pipe flow. As a whole, there are nine components that can contribute to the overall FATNAM system in the WGM stages.

The wet season is defined as the months of rainfall, and especially the periods of rainfall occurrences over each wadi. For this purpose, the wet and dry seasons can be identified on the basis of monthly rainfall amounts, which were calculated by Al-Sefry et al. (2005) and Es-Saeed, et al. (2005) for wadis Fatimah and Na'man, respectively, and are presented in Table 9.21. The calculation of the average monthly rainfall volumes is achieved on the basis of isohyetal maps for many years over each wadi. These are the volumes that are expected to recharge the alluvial deposits in each wadi. This table indicates that Wadi Fatimah is richer than Wadi Na'man. The annual total rainfall water volume for wadis Fatimah and Na'man are 775.66×10^6 m^3 and 1013.99×10^6 m^3, respectively. Hence, the FATNAM system has 1789.65×10^6 m^3 annual total rainfall water availability.

In the logical WGM procedure, the connections between the sub-wadi units are achieved by two logical conjunctions: "ANDing" or "ORing". If the sub-wadis are in serial connection, then all the serial components must function for the overall performance of the system, and therefore the units are connected by "ANDing". Whereas,

TABLE 9.21

Average Monthly Rainfall Volumes (×10⁶ m³)

Wadi						Months							
	Jan	Feb	Mar	Apr	May	Jun	Jul	Aug	Sep	Oct	Nov	Dec	Total
Fatimah	101.67	64.82	65.64	116.19	117.61	52.62	40.31	74.89	84.95	66.13	118.18	111	1013.99
Na'man	80.73	50.09	62.29	51.01	73.78	39.36	33.7	56.22	79.15	63.38	102.23	83.72	775.66
Total	182.4	114.91	127.93	167.2	191.39	91.98	74.01	131.11	164.1	129.51	220.41	194.7	1789.65

parallel units contribute to the overall performance of the system by the "ORing" connective. The statements, including "ANDing's and "ORing's, are referred to as a rule base." In this manner, it is possible to extract the relevant rules for each conceptual model under consideration. The WGM may require more than one rule and, in general, there are many rules. Each rule indicates an independent response of the system to the demand. Let us consider the wadi Fatimah configuration only, as presented in Figure 9.37. This leads to the rule bases as in the following, which provides an opportunity to assess the entire Fatimah basin sub-wadi units in an exhaustive management program:

1. (Ash-Shamiyah)AND(Bani Umayr)
   OR
2. (Ash-Yamaniyah)AND(Bani Umayr)
   OR
3. (Ash-Shamiyah)AND(Al-Yamaniyah)AND(Bani Umayr)
   OR
4. (Ash-Shamiyah)AND(Al-Yamaniyah)OR(Bani Umayr)
   OR
5. (Ash-Shamiyah)AND(Al-Yamaniyah)AND(Bani Umayr)AND(Alaf)
   OR
6. (Ash-Shamiyah)AND(Al-Yamaniyah)AND(Bani Umayr)OR(Alaf)
   OR
7. (Ash-Shamiyah)OR(Al-Yamaniyah)AND(Bani Umayr)AND(Alaf)
   OR
8. (Ash-Shamiyah)OR(Al-Yamaniyah)AND(Bani Umayr)OR(Alaf)

Similarly, considering the sub-unit configuration from Figure 9.37, the logical management rules for Wadi Na'man can be deduced as in the following:

1. (Majarish)AND(Na'man)
   OR
2. (Ash-Sharah)AND(Na'man)
   OR
3. (Ar'ar)AND(Na'man)
   OR
4. (Rahjan)AND(Na'man)
   OR
5. (Majarish)OR(Ash-Sharah)AND(Na'man)
   OR
6. (Majarish)OR(Ar'ar)AND(Na'man)
   OR
7. (Ash-Sharah)OR(Ar'ar)AND(Na'man)
   OR
8. (Majarish)AND(Ash-Sharah)AND(Ar'ar)AND(Na'man)
   OR
9. (Majarish)AND(Ash-Sharah)AND(Ar'ar)AND(Na'man)AND(Rahjan)

OR
10. (Majarish)AND(Ash-Sharah)AND(Ar'ar)AND(Na'man)OR(Rahjan)

Hence, for wadi Na'man within-basin management, there are ten independent logical management rules.

To supply groundwater from the FATNAM system, there are many independent rules, and in total there are $8 \times 10 = 80$ alternatives. It is not necessary to write down all these rules, but the most logical ones are listed in the following:

1. (Ash-Shamiyah)AND(Al-Yamaniyah)AND(Bani Umayr)AND(Alaf) AND(Majarish)AND(Ash-Sharah)AND(Ar'ar)AND(Na'man) AND(Rahjan)
   OR
2. (Ash-Shamiyah)AND(Al-Yamaniyah)AND(Bani Umayr)AND(Majarish) AND(Ash-Sharah)AND(Ar'ar)AND(Na'man)
   OR
3. (Ash-Shamiyah)AND(Al-Yamaniyah)AND(Bani Umayr)AND (Majarish) AND(Ash-Sharah)AND(Na'man)
   OR
4. (Ash-Shamiyah)AND(Bani Umayr)AND(Rahjan)AND(Na'man)
   OR
5. (Ash-Shamiyah)AND(Al-Yamaniyah)AND(BaniUmayr)AND(Alaf)AND (Rahjan)AND(Na'man)
   OR
5. (Ash-Shamiyah)AND(Al-Yamaniyah)AND(Bani Umayr)AND(Majarish) AND(Ash-Sharah)AND(Ar'ar)AND(Na'man)AND(Rahjan)
   OR
6. (Ash-Shamiyah)AND(Bani Umayr)AND(Majarish)AND(Na'man)
   OR
7. (Ash-Shamiyah)AND(Al-Yamaniyah)AND(Bani Umayr)AND(Majarish) AND(Ash-Sharah)AND(Ar'ar)AND(Na'man)
   OR
8. (Ash-Shamiyah)AND(Al-Yamaniyah)AND(Bani Umayr)AND(Alaf) AND(Majarish)AND(Ash-Sharah)AND(Ar'ar)AND(Na'man) AND(Rahjan)

It is a difficult task to identify the best and the most strategic joint WGM rule among these logical rules. However, as the specific aspects are taken into consideration, the picture becomes apparent (Şen, 2008).

REFERENCES

Al-Sefry, S., Şen, Z., Al-Ghamdi, S.A., Al-Ashi, W., and Al-Baradi, W. 2005. *Strategic ground water storage of Wadi Fatimah—Makkah region, Saudi Arabia.* Saudi Geological Survey, Hydrogeology Project Team, Final Report, 134 pp.

Altunkaynak, A., Özger, M., and Şen. 2003. Triple diagram model of lake level fluctuations in Lake Van, Turkey, *Hydrol. Earth Syst. Sci.* 7(2): 235–244.

Benjamin, J.R. and Cornell, C.A. 1973. *Probability Statistics and Decision Making in Civil Engineering,* McGraw-Hill, New York.

Cressie, N.A.C. 1993. *Statistics for Spatial Data, revised edition*, Wiley, New York, 900 pp.

ECI, European Cooperation in Informatics. 1978. Giampio Bracchi, Peter C. Lockemann (Eds.): Information Systems Methodology, *Proc. 2nd Conf. Eur. Cooperation in Informatics*, Venice, Italy, October 10–12, 1978. Lecture Notes in Computer Science 65 Springer 1978, ISBN 3-540-08934-9.

EPA, 1997. Policy for use of Monte Carlo analysis in risk assessment. Environmental Protection Agency, Washington DC, USA.

Esogbue, A.O., Theologidu, M., and Guo, K. 1992. On the application of fuzzy sets theory to the optimal flood control problem arising in water resources systems. *Fuzzy Sets and Systems* 48: 155–172.

Es-Saeed, M., Şen, Z., Basamad, A., and Dahlawi, A., 2005. *Strategic ground water storage of wadi Na'man—Makkah region, Saudi Arabia*, Saudi Geological Survey, Hydrogeology Team, Final Report, 118 pp.

Goulter, I.C. and Tai, F.-K. 1985. Practical impressions in the use of stochastic dynamic programming for reservoir operation. *Water Resources Bull.* 21 b(1): 65–74.

Isaaks, E.H. and Srivastava, R.M. 1989. *An Introduction to Applied Geostatistics.* Oxford University Press, Oxford, 561 pp.

Journel, A.G. and Huijbregts, C.I. 1978. *Mining Geostatistics.* Academic Press, London, 710 p.

Kadioglu, M. and Şen, Z. 2001. Monthly precipitation-runoff polygons and mean runoff coefficients. *Hydrolog. Sci. J.* 46(1): 3–11.

Karamouz M. and Houck, M.H. 1982. Annual and monthly reservoir operating rules, *Water Resour. Res.* 18(5): 1337–1344.

Kaufmann, A. and Gupta, M.M. 1985. *Introduction to Fuzzy Arithmetic.* Van Nostrand Reinhold, New York.

Kiska, J., Gupla, M., and Nikiforuk, P. 1985. Energetic stability of fuzzy dynamic systems. *IEEE Trans. Systems, Man and Cybern.,* 15: 783–792.

Kosko, B. 1987. Fuzzy associative memories. In *Fuzzy Expert Systems*, Kandel, A., Ed. CRC Press, Boca Raton, FL.

Kundu, B., Richardson, S.D. Swartz, P.D. Matthews, P.P. Richard, A.M., and Demarini, D.M. 2004. Mutagenicity in salmonella of halonitromethanes: a recently recognized class of disinfection by-product in drinking water. *Mutat. Res.* 562: 39–65.

Loucks, D.P. et al. 1981. *Water Resources Systems Planning and Analysis.* Prentice-Hall, Englewood Cliffs, NJ.

Mamdani, E.H. 1974. Application of fuzzy algorithms for simple dynamic plant. *Proc. IEE,* 121: 1585–1588.

Matheron, G. 1963. Principles of geostatistics. *Economic Geol.* 58: 1246–1266.

Matheron, G. 1965. *Les Variables Regionalisees et Leur Estimation, Masson*, Paris, 306p.

Matheron, G. 1971. *The Theory of Regionalized Variables and its Applications.* Ecole de Mines, Fontainbleau, France.

Myers, D.E., Begovich, C.L., Butz, T.R., and Kane, V.E. 1982. Variogram models for regional groundwater chemical data. *Math. Geol.* 14: 629–644.

Rajkumar, T., Guesgen, H. W., and Gorman, D. 1999. Estimating the consumption of tap water using fuzzy concepts. Proc. Int. ICSC Cong., Computational Intelligence Methods and Application, Rochester, NY. USA, 1–5.

Raucher, R.S., Frey, M.M., and Cook, P.L. 2000. *Benefit-cost analysis and decision-making under risk uncertainty: issues and illustrations.* Interdisciplinary Perspectives on Drinking Water Assessment and Management. IAHS Publ. No. 260, 141-70.

Ross, J.T. 1995. *Fuzzy Logic with Engineering Applications.* McGraw-Hill, New York, 593 pp.

Şen, Z. 1996. *Yildiz derelerinin—Durusu (Terkos) gölü ile ortak çalistirilmasi*, (Joint operation of Yildiz creek dams with Durusu (Terkos) lake. Istanbul Water and Sewerage Administration, Final Report, 156 pp.

Şen, Z. 2001. *Bulanık Mantık ve Modelleme Ilkeleri (Fuzzy Logic and Modelling Principles)*, Bilgi, Sanat ve Kültür Basimevi, Istanbul (in Turkish).

Şen, Z. 2005. Strategic groundwater resources planning for Makkah Al-Mukharramah. FATNAM system. Saudi Geological Survey (in preparation).

Şen, Z. 2008. Wadi Hydrology. Taylor and Francis Group, CRC Press, Boca Raton, 347 pp.

Şen, Z. 2009. *Spatial Modeling Principles in Earth Sciences*. Springer Verlag, (in print).

Suharyanto, C.X., and Goulter, I.C. 2003. Resservoir operating rules with fuzzy programing. *J Hydraul. Engrg.*, ASCE.

Thompson, M.A., Kyle, R.A., Melton, L.J. III., Plevak, M.F., Rajkumar, S.V. 2004. Effect of statins, smoking and obesity on progression of monoclonal gammopathy of undetermined significance: a case-control study. *Haematologica* 89: 626–628.

USEPA. 1996. *Exposure Factors Handbook.* U.S. Environmental Protection Agency, Office of Research and Development National Center for Environmental Assessment, Washington, D.C.

Zadeh, L.A. 1965. *Fuzzy Sets, Information Control,* 8, 338–353.

Index

Printed and bound by CPI Group (UK) Ltd, Croydon, CR0 4YY

18/10/2024

01776243-0011